# Innovative Cultural Tourism
# in European Peripheries

Cultural tourism can play an important role in social and territorial cohesion. Focusing on European peripheral regions, this book illuminates the importance of local communities in heritage management for sustainable development.

This book provides insights into the use of innovative business models and tools, such as ecosystem services contracts and digital narrative platforms, to enhance the sustainability and economic development of peripheral and marginal destinations. Additionally, this book addresses the value of data collection and analysis in cultural tourism and provides insights into participatory models and approaches that contribute to sustainable tourism development.

With contributions from a pan-European range of expert scholars and practitioners, this book serves as an essential resource for researchers, professionals, and anyone with an interest in tourism, and the cultural and creative industries.

**Karol Jan Borowiecki** is Professor of Economics at the University of Southern Denmark, and President-Elect of the Association for Cultural Economics International.

**Antonella Fresa** is Director of Implementations at Promoter srl, Vice-President of Photoconsortium: international association and contracted professor of Tourism Science at the University of Pisa, Italy.

**José María Martín Civantos** is Professor of Medieval History and Archaeology at the University of Granada, Spain.

# Innovative Cultural Tourism in European Peripheries

Edited by
Karol Jan Borowiecki,
Antonella Fresa and
José María Martín Civantos

Routledge
Taylor & Francis Group

LONDON AND NEW YORK

First published 2025
by Routledge
4 Park Square, Milton Park, Abingdon, Oxon OX14 4RN

and by Routledge
605 Third Avenue, New York, NY 10158

*Routledge is an imprint of the Taylor & Francis Group, an informa business*



*British Library Cataloguing-in-Publication Data*
A catalogue record for this book is available from the British Library

ISBN: 9781032728995 (hbk)
ISBN: 9781032729046 (pbk)
ISBN: 9781003422952 (ebk)

DOI: 10.4324/9781003422952

Typeset in Sabon
by codeMantra

# Contents

# Figures

# Tables

# Contributors

**Dora Agapito** holds a PhD in Tourism. She is Assistant Professor at the Faculty of Economics, University of Algarve, Portugal. She is Director of the PhD in Tourism and the MSc in Tourism Organizations Management, and Researcher at CinTurs—Research Center for Tourism, Sustainability and Well-Being.

**Shahedul Alam Khan** is a postgraduate student from Bangladesh pursuing his MSc in Economics and Business Administration at the University of Southern Denmark. Besides, he is working as a student research assistant at the Department of Economics. Before his academic pursuits in Denmark, he held the position of Assistant Professor at Leading University in Bangladesh. His research interests span governance, international trade, and business regulation, and he has several publications to his credit.

**Desidério Batista** holds a PhD in Landscape Arts and Techniques. He is Professor at the University of Algarve on the MSc in Landscape Architecture and the PhD in Heritage Studies. He is a researcher at the Centre for Studies in Archaeology, Arts and Heritage Sciences (CEAACP).

**Alexandra Bitušíková** is Head of the University Centre for International Projects and Full Professor in Social Anthropology at Matej Bel University in Banská Bystrica, Slovakia. She was Visiting Scholar at Cambridge University, University College London, and Boston University (Fulbright). She participated in numerous FP, H2020, and HEU projects and is an author of more than 100 publications on urban change, diversity, identity, heritage, and gender. From 2001 to 2008, she worked in the European Commission and the European University Association in Brussels.

**Maximilian Block** holds a Bachelor's degree in International Tourism Management and a Master's degree in African Studies. He is currently about to graduate with a Master of Science in Sustainable Entrepreneurship from the University of Groningen and is concurrently working as Student Research Assistant for Copenhagen Business School. His research interests involve interdisciplinary approaches to sustainable development with a key focus on Sub-Saharan Africa, cultural tourism development, and nature-based carbon offsetting.

**María Teresa Bonet García**, Professional Archaeologist since 2010, has participated in several research projects in collaboration with the University of Granada, mostly related to studies linked to Landscape Archaeology. She has actively participated as a partner in the development and execution of the Mediterranean Mountainous Landscapes project (MEMOLA Project, FP7, European Commission, coordinated by the University of Granada). She is currently part of the management and support team of the INCULTUM project. She is specialized in Hydraulic and Territorial Archaeology, with a good command of Geographic Information Systems (GIS).

**Kamila Borseková** is Coordinator of Research at Matej Bel University with a profound interest in urban and regional topics. She is Associate Professor and Head of the Research and Innovation Centre at the Faculty of Economics. Her primary scientific research focuses on competitiveness, resilience, and sustainable urban and regional development and related policy. She has extensive experience from participating in dozens of national and international research projects. Currently, she is Coordinator of the Horizon Europe project BRRIDGE. She has authored and co-authored more than 100 scientific publications, including articles, chapters, studies, and books.

**Karol Jan Borowiecki** is Professor of Economics at the University of Southern Denmark, renowned for his innovative research methodologies and societal impact. He published more than 40 items, including in the *Journal of Political Economy*, a textbook with Cambridge University Press, and a co-edited volume on cultural heritage. He sits on the editorial boards of *Tourism Economics* and the *Journal of Cultural Economics*. He serves as President-Elect of the Association for Cultural Economics International and collaborates with premier European institutions, shaping policy and advancing cultural and tourism economics.

**Marisa Cesário** holds a PhD in Economics. She is Professor at the Faculty of Economics, UALG (Portugal). She is Director of the MSc in Tourism Economics and Regional Development. She is an integrated member of CinTurs—Research Center for Tourism, Sustainability and Well-Being.

**Flore Coppin** is Head of communications at Bibracte and Coordinator of the INCULTUM pilot project. She graduated from the National Institute for Oriental Languages and Civilizations and the Sorbonne School of Communication.

**Elena Correa Jiménez**, archaeologist, graduated in Archaeology from the University of Granada in 2020 and Master's degree in Archaeology from the University of Granada in 2021. She is specialized in Historical Irrigation Systems in the southeast of the Iberian Peninsula. She is in charge of the volunteer activities of the Biocultural Archaeology Laboratory (MEMO-Lab) such as the recovery of irrigation ditches and historical cultivation areas. She has also participated in numerous archeological excavations in the province of Granada and Almería. She is currently Researcher on the project, Visiting the Margins: INnovative CULtural ToUrisM in European peripheries (INCULTUM).

**Miguel Reimão Costa** is an architect and professor at the University of the Algarve, a researcher at the Centre for Studies in Archaeology, Arts and Heritage Sciences (CEAACP), and a member of the board of the Mértola Archaeological Site (CAM).

**Antonella Fresa** is Director of implementations at Promoter srl, Contracted Professor of Tourism Science at the University of Pisa, and Vice-President of Photoconsortium: international association for valuing photographic heritage and accredited aggregator of Europeana.eu. She holds a Master's degree in Computer Science.

**Mikael Gidhagen** is Senior Lecturer in Marketing at the Department of Business Studies at Uppsala University, Sweden. His research concerns aspects of service logic, including actor engagement and value creation in service ecosystems. Focal interests are relationship dynamics, engagement, experienced value, and resource becoming.

**Manuela Guerreiro** holds a PhD in Economic and Management Sciences. She is Programme Leader of the Master of Marketing Management and Co-Coordinator of the Research Centre for Tourism, Sustainability and Well-Being (CinTurs). Her research interests include Marketing and Consumer Behavior, Destination Branding, and Image.

**Vincent Guichard** is General Manager of Bibracte Public Establishment for Cultural Cooperation. He is an archaeologist specializing in European protohistory. His work and responsibilities are focused on preserving and developing the Bibracte archeological site, while maintaining the establishment as a key public player in the fields of archeological research and integrated management of landscapes and cultural heritage.

**Eszter György** received her MA at Eötvös Loránd University (ELTE) in Budapest, at the EHESS, Paris, and her PhD in History at ELTE, in 2013. Since 2012, she has been involved in several EU projects (Erasmus Mundus and Horizon 2020 projects), focusing on cultural heritage, cultural participation, and minority heritage. She is a senior lecturer at the Atelier Department for Interdisciplinary History, ELTE. Her fields of research and publications cover Roma cultural history, Roma heritage in Hungary, and urban inequalities.

**Carsten Jacob Humlebæk** is Associate Professor at Copenhagen Business School, Denmark. His research focuses particularly on questions related to collective identity formation as a social and narrative phenomenon and on the dialectical relationship between place branding and collective identity building both at national, regional, and local levels. His research has been published in numerous academic journals and books. He has been involved in several international research projects and is presently CBS-lead of the EU-funded "INCULTUM" and "SECreTour" projects.

**Vaios Kotsios (Greek)**, PhD (2016) in Environmental and Development Sciences from the National Technical University of Athens, specializes in Business Intelligence, GIS, and Data Analysis. Since 2018, he's lectured in NTUA's "Environment and Development" program, previously researching there since 2009. As a chief researcher and data analyst, he contributed to the Mechanism of Labour Market Diagnosis by the Greek Ministry of Labour. From 2018 to 2020, he served as a national expert in the European Social Fund's Transnational Network for Employment. He's collaborated with various institutions nationally and internationally, including the UN, EU, and Greek ministries.

**José María Martín Civantos** is Full Professor in the Department of Medieval History and CCTTHH at the University of Granada, Coordinator of Biocultural Archaeology Laboratory (MEMOLab)—coordinator and IP of the successful and highly awarded European project MEMOLA (MEditerranean MOntainous LAndscapes: an historical approach to cultural heritage based on traditional agrosystems), funded by the EU, under FP7/SSH. Also, he is Coordinator of INCULTUM Project (financed by the H2020 program of the European Union under Grant Agreement no. 101004552). He is an archaeologist specializing in landscape studies, local community relations, and archaeology of Architecture.

**Sara Beth Mitchell** specializes in urban, labor, and cultural economics, focusing on creative worker migration, agglomeration effects, and cultural tourism. Her work leverages often historical datasets, shedding light on economic patterns. Holding a PhD from Trinity College Dublin, where she was a Grattan Scholar and Irish Research Council Fellow, her career spans roles at the University of Southern Denmark, TU Dortmund, and the Institute of Public Administration, reflecting her diverse expertise and contributions to economics.

**Ardit Miti** is a researcher at the Centre for the Research and Promotion of Historical-Archaeological Albanian Landscapes, based in Tirana, Albania. He is also a PhD candidate at the University of Granada, Spain, focusing his research on landscape archaeology and rural settlement dynamics and patterns during Late Antiquity and the Early Middle Ages.

**Anna-Carin Nordvall** is Associate Professor at the Department of business studies, Uppsala University. Her research focuses on individual aspects of the decision-making process in different contexts, for example consumers, visitors, and organizational strategists. In her work, she consolidates empirical quantitative and qualitative data modeling human cognition and perception for sustainable development.

**Gábor Oláh** holds a PhD in Urban Studies and History and is Postdoctoral Researcher at ELTE Eötvös Loránd University, Budapest and Paris 1 Panthéon-Sorbonne University. In 2023, he defended his PhD dissertation in the framework of ELTE-EHESS joint supervision program. His research is focused on urban heritage discourse, the concepts of urban landscape and neighborhood, as well as issues related to culture-led urban regeneration. He has been involved in several EU-funded and other international research projects: REACH (2017–2020), UNCHARTED (2020–2024), SECreTour (2024–2027), and HerEntrep (2024–2027).

**John Östh** is Swedish Professor and Geographer with a special interest in studies of human mobility, urban and regional planning, and methods development. With a descent from one of the larger Swedish archipelagos, he has an interest in regional development of rural and coastal areas. Using data from GPS, mobile phones, and map repositories, he has contributed to knowledge generation from the more quantitative perspective. John holds a chair in urban planning at Oslomet University, Norway and a guest professorship at Uppsala University.

**Esben Rahbek Gjerdrum Pedersen (ERGP)** is Professor at Copenhagen Business School, Denmark. His research focuses on business model innovation and the operationalization of new management ideas, including corporate sustainability, non-financial performance measurement, and lean management. His research has been published in numerous academic journals and received international recognition, including Emerald Outstanding Paper Award and Emerald Social Impact Award. ERGP has also co-developed two massive open online courses (MOOCs) on Sustainable Fashion and Business Models for Sustainability (available on Coursera).

**Maja Uhre Pedersen** holds a PhD degree in Economics, is currently Assistant Professor of Economics at the University of Southern Denmark, has specialized in Economic History and Growth with an emphasis on data analysis, and has published six items on different topics, including economic growth, financial history, and globalization—all published in various peer-reviewed journals.

**Sabine Gebert Persson** is Associate Professor in Marketing at the Department of Business Studies, Uppsala University, Sweden. Her current research explores complexities in actor engagement in a spatiotemporal context, contributing to knowledge of how interactions between the visitor, the visited, and the place, shape sustainable destination development. With a special interest in how interactions form behavior, her research includes network organizations and legitimacy formation.

**Darina Rojíková** is a researcher at the Research and Innovation Centre, Faculty of Economics, Matej Bel University, in Banská Bystrica, Slovakia. She has an interest in urban and regional topics. Her primary scientific research encompasses urban and regional development, the territorial development of peripheral areas, strategic planning of territorial development, democracy, and public policy. Rojíková obtained her professional qualifications through subsequent studies at the Faculty of Economics at Matej Bel University. She has been involved in several research projects and has authored and co-authored more than 25 scientific publications.

**Bernardete Sequeira** holds a PhD in Sociology. She is Professor in the Faculty of Economics, University of Algarve. She is a researcher of the Interdisciplinary Centre of Social Sciences—CICS.NOVA. Her research interests include Sociology of Organizations and Work, Knowledge Management, and Tourism.

**Eglantina Serjani** is a researcher at the Centre for the Research and Promotion of Historical-Archaeological Albanian Landscapes, based in Tirana, Albania. She is an archaeologist specializing in the Late Roman archeology and in the digital management and mapping of archeological and heritage data.

**Amanda Slattery** is Development Manager with Ballyhoura Development. Amanda focuses on the development, coordination, and implementation of broad-based community and locally led economic development strategies and initiatives across all of Ballyhoura Development's programs and projects. Amanda has been instrumental in the growth of tourism and recreation across the Ballyhoura area and the increase in visitor numbers over time in her role as Tourism and Heritage Officer. Key to the approach has been the generation of innovative and creative content for visitors to engage with online and innovative community-led heritage and tourism-based initiatives. Amanda has led initiatives such as place-based living and learning, audio visual content generation program that has built capacity for people across the Ballyhoura area to share their stories; Ballyhoura Reaching out; Community-led genealogy programs and Awareness-Raising Initiatives for Social Enterprises within the Ballyhoura Region of North Cork and East Limerick. Amanda has expertise in project management and in leading and facilitating multi-agency projects.

**Viktor Smith** is a PhD in International Business Communication and an Associate Professor at the Department of Management, Society and Communication, Copenhagen Business School. His key research interests are innovative naming and framing processes and multimodal communication applied to fields such as product packaging design, communicative fairness, sensory presentation of foods, sustainable tourism development, and place branding. He has led and participated in a number of cross-institutional R&D projects and published a number of scientific studies and popular science works.

**Gábor Sonkoly** (PhD EHESS, Paris, 2000; Doctor of the Hungarian Academy of Sciences, 2017) is a Professor of History at the Atelier Department for Interdisciplinary History and Head of the History PhD School at ELTE University, Budapest. He published/edited 20 books in English, French, Hungarian, Japanese, and Portuguese, as well as 100 articles/chapters on urban history, urban heritage, and critical history of cultural heritage. He presented at 120+ international colloquia and was a guest professor in 15 countries. He is Chair of the Panel for European Heritage Label since 2020 and an EU expert since 2013.

**John Tierney** is a field archeologist based in Ireland. As Senior Archaeologist from 1999, he focused on excavation and publication. Since 2010, he has been Director of the Historic Graves Project combining community graveyard surveys with the use of non-invasive methods of unmarked grave identification and recording.

**Marina Toger** is Senior Lecturer at the Department of Human Geography at Uppsala University, Sweden. Her research focuses on quantitative spatiotemporal analysis using geocomputation, GIS, and agent-based modeling, particularly concerning urban and regional mobility and social inequality, and applying the complex systems approach. Her work integrates empirical data analysis and theoretical modeling, fostering international collaborations, and interdisciplinary projects aimed at sustainable regional futures.

**Maurizio Toscano** (PhD) is an ICT, Web Information Systems, and Data Management specialist, based in Spain. As a researcher, he works in the field of Digital Humanities and Cultural Heritage, and he has over a decade of experience as a project manager in international collaborative projects.

**Sotiris Tsoukarelis (Greek)** is Founder and President of The High Mountains Social Cooperative Enterprise. With a background in political science, specializing in political philosophy and local development, he studied at the Metsovio Centre for Interdisciplinary Research in 2012. Since then, he's been an advocate for uplands, decentralization, and the agri-food sector. Tsoukarelis actively farms vegetables at 1000 meters altitude in Demati Zagori. He's trained in Tourism Experience Creation and Agrotourism by the American Farm School and is a Dairy School of Ioannina graduate. Additionally, he heads the Union of Social and Solidarity Economy Organizations of Epirus.

**Katarína Vitálišová** is Associate Professor at the Department of Public Economics and Regional Development, Faculty of Economics, at Matej Bel University. Her research focuses on public and participatory governance, strategic planning in spatial development, and the implementation of innovative approaches, such as those pertaining to creative and smart cities.

# Acknowledgments

The editors would like to thank Naomi Round Cahalin for her help with editing the manuscript and Maria Teresa Bonet for efficiently managing the editorial process.

This book was developed as part of the innovation action under the project INCULTUM—Visiting the Margins: Innovative Cultural Tourism in European peripheries, funded by the EU's Horizon 2020 program (grant agreement no 101004552). We are thankful to Rodrigo Martin Galan, Project Officer, for his advice and support. More information about INCULTUM can be found on the project's website at www.incultum.eu.

# 1 Introduction

## Visiting the margins: innovative cultural tourism in European peripheries

*Karol Jan Borowiecki, José María Martín Civantos and Antonella Fresa*

Peripheral and marginalized areas possess untapped potential for developing sustainable tourism. Often overshadowed by more prominent destinations, these regions are rich in unique cultural and natural heritage that can significantly contribute to social and territorial cohesion. Sustainable cultural tourism emerges as a strategic avenue to harness these assets, not merely for economic benefits but as a means to reinforce cultural identity, community engagement, and environmental conservation (Richards, 2018).

Moving away from traditional, consumption-driven models, the development of cultural tourism in these regions advocates for an inclusion and sustainable paradigm. This shift emphasizes cultural tourism's role as a driving force behind local development, improving the recognition of marginalized areas, and ensuring a more equitable share of tourism's socioeconomic benefits. These perspectives align with the discussions in Borowiecki et al. (2016), which recalibrates the relationship between institutional and individual cultural heritage practices in the face of 21st-century changes.

Building upon this premise, this book delves into sustainable cultural tourism and its significant role in social and territorial cohesion. It showcases innovative solutions and strategies developed by the EU-funded Horizon 2020 project, INCULTUM, an Innovation Action aimed at promoting sustainable cultural tourism in Europe's remote and marginal areas through collaborative and participatory approaches. The project, titled "Visiting the Margins: Innovative Cultural Tourism in European Peripheries," spanned three years, from May 2020 to April 2024, displaying a commitment to transforming the landscape of cultural tourism through innovative practices.

The initiative brought together 15 partners from across ten European countries, including Denmark, France, Greece, Ireland, Italy, Portugal, Slovakia, Spain, Sweden, and Albania. The large consortium was designed to cover a broad spectrum of skills, expertise, and the entire value chain necessary to achieve the project's primary goals effectively, thus maximizing its impact across the EU. Furthermore, it embraced a diverse array of European peripheries, showcasing the vast and varied cultural landscape of the continent's lesser-known regions. It aimed to engage a comprehensive array of

DOI: 10.4324/9781003422952-1

stakeholders, including cultural tourism organizations, local interest groups, communities, parks, municipalities, associations, universities, small- and medium-sized enterprises, data managers, developers, and networks among EU cities, as well as development and consulting firms. Several consortium partners initiated pilot studies, integrating innovation efforts with experimental adjustments at selected locations to evaluate and enhance sustainable tourism practices. Within the scope of the pilots and beyond, the consortium undertook both practical and theoretical work, exploring participation models, mapping out stakeholders, and experimenting with data usage.

The ambition of INCULTUM was to transform the concept of cultural tourism from merely a consumer product into a social tool. INCULTUM's innovation lies in unlocking the potential of cultural tourism to benefit local communities and stakeholders, aiming to enhance social, cultural, and economic development sustainability. Tourism should be much more than consuming, and the project aimed to demonstrate that visitors, recipient territories, and communities can benefit from this activity. Throughout the duration of the project, three key strategies were explored: enhancing governance through public participation, diversifying and implementing solutions based on the circular economy, and conducting ongoing monitoring and evaluation to improve the adaptability and effectiveness of the actions proposed. This innovative approach to cultural tourism is designed to empower host territories with strategies to manage, reduce, or redistribute tourist flows more effectively. INCULTUM has showcased this through a blend of activities, including the implementation of digital tools, fostering networks among stakeholders with shared interests, and demonstrating ways to meet expectations. Importantly, this novel approach to cultural tourism also critically considers the impacts of current global changes.

This book highlights the importance of local communities in the management of heritage resources and their significant contribution to the development of cultural and sustainable tourism initiatives. It also examines the creation of innovative itineraries that highlight shared heritage resources and incorporate local involvement. The text emphasizes the adoption of innovative business models and tools, such as contracts for ecosystem services and platforms for digital storytelling, to improve both sustainability and economic growth in peripheral and marginal destinations. Additionally, it explores the critical role of data collection and analysis in cultural tourism, offering insights into participatory models and methods that support sustainable tourism development. This book also delves into territorial entrepreneurship and the comprehensive approach necessary to foster socio-economic growth while preserving the quality of the landscape and heritage.

Readers will gain insight into innovative strategies and practices for sustainable cultural tourism; they will understand the crucial role of local communities in heritage management; discover business models and tools designed to foster economic development in peripheral and marginal destinations; acquire techniques for effective data collection and analysis aimed at cultural tourism planning and assessment; become familiar with participatory models and methods that promote sustainable tourism while aligning with community values; and comprehend the role of territorial entrepreneurship in socio-economic development and the preservation of heritage.

This book is structured into two distinct yet complementary parts, spanning 15 chapters, including this Introduction. The first part delves into theoretical approaches and cross-cutting themes such as innovation, economics and business models, data collection and analysis, participatory models, and the interplay between identity and cultural model. The second part focuses on eight pilot cases, illustrating applied research and innovation efforts.

The first part of the book lays the foundation with its initial focus on the project's central vision in Chapter 2, titled "Tourism as a Tool for Social and Territorial Cohesion – Exploring the Innovative Solutions Developed by INCULTUM Pilots". This chapter highlights the innovative approaches pursued by the INCULTUM pilots to promote sustainable cultural tourism, underscoring the importance of community involvement and the application of business intelligence tools to create visitor itineraries that enhance project sustainability.

Chapter 3, "Place Branding from Scratch: Naming, Framing, and Finding", shifts the perspective to economic and entrepreneurial aspects. It examines the intricacies of bottom-up place branding, discussing the pivotal role of local engagement and cost-effective branding strategies in attracting tourists while preserving local identity and cultural heritage.

In Chapter 4, titled "Innovative Business Models for Cultural Tourism: Advancing Development in Peripheral Locations", the narrative continues with an analysis of peripheral and marginal destinations. It investigates collaborative business models that support economic, social, and cultural development in cultural tourism. This chapter reviews existing alternative models and proposes payment-for-service contracts as a means to alleviate negative impacts and redistribute benefits equitably.

Chapter 5 expands on the theme of "Data Collection and Analysis in Cultural Tourism". Highlighting the crucial role of data within cultural heritage institutions, it offers insights into effective data collection methods, including visitor surveys and data scraping algorithms. Additionally, it explores the concept of data-driven storytelling and provides practical tools for data management, monitoring, and evaluation in cultural tourism.

The discussion then progresses to participatory approaches in Chapter 6, "Participatory Models and Approaches in Sustainable Cultural Tourism", emphasizing their significance in sustainable tourism and culture development. This chapter explores the overlap between participation, innovation, and digitalization, proposing a novel participatory framework for sustainable cultural tourism initiatives.

Concluding the first part, Chapter 7, "Identity and Cultural Impact of Sustainable Cultural Tourism", explores the importance of preserving cultural heritage and encouraging the promotion of local traditions. This chapter illustrates the transformative power of cultural tourism in fostering a sense of identity and pride within communities and among visitors, thereby enriching the cultural tourism experience and fostering deeper connections to place.

The second part of the book presents eight pilot projects developed across Europe, outlined in Chapters 8–15. These projects involve partners, stakeholders, and communities from eight different countries, encompassing a diverse range of geographical settings. Emphasis has been placed on remote areas along with their cultural and natural heritage. The pilots have focused on various types of resources: industrial heritage in Slovakia's mining regions; agrarian heritage in Portugal and Spain; archaeological sites in France, Spain; the natural landscapes of Albania, Greece, France, Spain, and in the Swedish archipelagos; the cultural practices of minorities in Albania and Greece; and the local identities found in Ireland's historic graveyards and the mountain villages of Western Greece. These diverse locations narrate the rich histories of their inhabitants, and their ongoing efforts to leverage their regional potential, drawing visitors with the allure of authentic traditions.

Chapter 8 introduces the first pilot, "The Mining Treasures of Central Slovakia," detailing the development of a participatory platform for sustainable cultural tourism in the region. This platform connects the Barbora Route with the European Fogger Route, highlighting the area's rich mining heritage. The narrative emphasizes a gap in marketing and digital resources for promoting the Banská Bystrica region's mining history. Designed, tested, and marketed through a participatory approach involving researchers, students, stakeholders, and the public, the platform's potential impact on cultural tourism, supported by data analysis and community involvement, is thoroughly explored.

In Chapter 9, the focus turns to agrarian heritage in the Coastal Plain of the Algarve in Portugal, discussing its potential for community-based cultural tourism. This chapter features the cultural landscape shaped by water management in the Campina de Faro region, discussing the historical and cultural value of abandoned hydraulic structures and traditional irrigation practices. It demonstrates innovative strategies to promote sustainable cultural tourism based on water heritage.

Continuing the focus on agricultural heritage, Chapter 10 explores the rural traditions of the Altiplano de Granada in Spain. The pilot aims to revalue historical water management systems and promote their conservation. This chapter discusses the transdisciplinary approach to appreciating the cultural and environmental values of these irrigation systems and their traditional ecological knowledge, highlighting tourism's role in recognizing and diversifying economic activities linked to these systems' cultural and environmental benefits.

Chapter 11 shifts the focus to France for an integrated approach to territorial entrepreneurship in the Bibracte – Morvan des Sommets region. It emphasizes the involvement of local communities and elected representatives in shaping the project's territory and strategic planning. This chapter discusses the role of the management team in facilitating territorial entrepreneurship, fostering cooperation between different sectors, and promoting the sustainable socio-economic development of the region.

The geographical and thematic focus changes in Chapter 12, which focuses on rural tourist behaviour and engagement in Sweden. It discusses the challenge of attracting tourists while preserving cultural heritage in rural settings, presenting a method that combines GPS data, open street maps, and surveys to capture tourists' spatiotemporal engagement and its impact on perceived value.

In Chapter 13, the Greek pilot case explores tourism as a vehicle for learning and cross-cultural communication through the "Aoos, the Shared River" initiative. It focuses on the Worth-Living Integrated Development, highlighting innovative methods and technologies to understand the area's cultural context and the importance of community collaboration in cultural tourism development, including the use of Commons-Based Peer Production and Social Economy models to foster innovative cultural products and services.

The narrative then explores in Chapter 14 the Vlach minority heritage in the Upper Vjosa Valley, Southeast Albania. This chapter examines the historical significance of the Vlach ethnic group and their seasonal migrations, stressing the importance of preserving and showcasing the Vlach minority's cultural values through experimental archaeology, cultural memory, and tourism.

Finally, Chapter 15 transitions to a different setting and subject, presenting a grassroots-led tourism case study through the lens of community genealogy in Ireland's Historic Graves Project. This project's approach to creating a grassroots tourism product using geolocated surveys of historic graveyards demonstrates the impact on tourism, particularly among the Irish diaspora, and demonstrates the potential of genealogical tourism to extend the tourism season.

The pilots promoted communities of practice and had a beneficial impact on local communities socially, culturally, environmentally, and economically. By focusing on cultural tourism that leverages the unique aspects of living territories and communities, these initiatives helped mitigate the negative impacts of tourism through targeted training and by reinforcing local identities and social connections. Bottom-up local strategies aimed at sustainable cultural tourism focused on hidden potentials in remote, peripheral, or deindustrialized areas that traditional tourism often overlooks. The outcomes of these participatory approaches laid the groundwork for the co-creation of innovative tools, assessing the prerequisites for their broader application and expansion beyond the timeframe of the EU's funding period.

The pilots carried out place-based participatory approaches and specific strategies across diverse geographical, social, and cultural landscapes to enhance cultural tourism with fresh perspectives, social innovation, and collaboration among a broad and varied network of stakeholders. Local communities took centre stage, encompassing various sectors and giving particular emphasis to women, youth, and minorities. These initiatives provided deeper insights into the trends and potential of sustainable tourism, leveraging local strengths and capacities. Innovative solutions were explored and validated, ensuring that the knowledge gained could be replicated and adapted in future endeavours. Consequently, the potential for development extends beyond the project's pilots, since the insights and networks formed are accessible to other stakeholders and regions as well.

This book captivates with its array of geographical, disciplinary, and thematic perspectives, diverging to varying degrees from the mainstream, commercialized form of tourism. It redefines the notion of periphery, not just in terms of geography, but as a concept encompassing social and heritage aspects – revealing areas and sectors often overlooked or negatively impacted by tourism, rather than benefiting from it. The discussion on the relative nature of touristification and mass tourism, emphasizing that it is not only about high visitor numbers, adds depth to the understanding of these issues.

The journey of exploration and innovation does not conclude with the project's end. The real challenge lies in preserving the mobilized expertise beyond the lifecycle of INCULTUM's funding. The realms of data analysis, business models, participatory engagement, heritage and territorial protection, community dynamics, equitable distribution of tourism benefits, local cohesion, and governance require continued attention. These efforts aim to mitigate the negative effects of tourism and foster a European dialogue on these critical themes, sharing best practices and learned lessons.

This book concludes with reflective thoughts on viewing tourism as a tool rather than the ultimate goal, underscoring the importance of developing inclusive and sustainable tourism projects. This involves establishing a heritage community united by common values, as inspired by Elinor Ostrom's pioneering work, and supported by the Council of Europe's Framework Convention on the Value of Cultural Heritage for Society (Faro Convention, 2005), advocating for heritage to be seen as a collective good managed by communities for its enduring preservations.

## Bibliography


Borowiecki, K. J., Forbes, N., & Fresa, A. (Eds.) (2016). *Cultural Heritage in a Changing World*. Cham: Springer. https://doi.org/10.1007/978-3-319-29544-2

Council of Europe. (2005). *Council of Europe Framework Convention on the Value of Cultural Heritage for Society*. Faro, Portugal: Council of Europe Treaty Series - No. 199.

Ostrom, E. (1990). *Governing the Commons: The Evolution of Institutions for Collective Action*. Cambridge: Cambridge University Press.

Richards, G. (2018). *Cultural tourism: A review of recent research and trends*. Journal of Hospitality and Tourism Management, 36, 12–21. https://doi.org/10.1016/j.jhtm.2018.03.005

# 2 Tourism as a tool for social and territorial cohesion

## Exploring the innovative solutions developed by INCULTUM pilots

*Flore Coppin and Vincent Guichard*

## 2.1 Introduction

What do a rural path in Burgundy, a historic irrigation system in Andalusia, an old Irish cemetery or a forgotten transhumance track in the mountains of Albania have in common? These are heritage resources that are highly valued by local communities.

This chapter summarises the solutions and methods explored by the INCULTUM pilot projects to develop sustainable cultural tourism based on cooperation and participatory approaches, using the attachment of local communities to their heritage as a lever. The perspective shared by these experiments is to consider tourism as a tool of strengthening the resilience of local communities and their living environment. The focus on the living environment, with all its biotic (or "natural") components, is of utmost priority as each passing year reveals the increasing impact of climate change, marking the onset of the Anthropocene. This change is not characterised by a momentary crisis but rather signifies the beginning of a new era of permanent tensions, necessitating a state of "prolonged urgency" (Kunstler, 2005).

The pilot projects encompass various complementary approaches that can be implemented sequentially and cyclically through an "innovation wheel" designed to initiate a virtuous circle. This innovation framework is based on the concepts of territorial common goods and heritage communities, the definition of which should be briefly recalled. The characteristics of the ten pilot projects will then be summarised, from the point of view of the commons to which they refer. The four stages of the process will then be discussed, with details of the tools used by INCULTUM to get through each stage.

## 2.2 A few reminders about territorial common goods and territorial heritage communities

The hypothesis developed in this chapter is that an essential condition for developing an inclusive and sustainable territorial tourism project is the establishment of a heritage community based on shared attachment reasons. To this end, heritage is considered in terms of its capacity to become a common

DOI: 10.4324/9781003422952-2

good, in the sense in which it has been understood since the pioneering work of Elinor Ostrom (1990), i.e. a resource that arouses a shared attachment among a group of people who manage it jointly by means of shared rules with a view to guaranteeing its long-term survival. This group of people thus constitutes a heritage community. This point of view on heritage is the one defended by the Council of Europe's Framework Convention on the Value of Cultural Heritage for Society (Faro Convention, 2005), which recognises the right of every person to benefit from cultural heritage and to contribute to its enrichment, in other words the right to designate what constitutes one's heritage, to take part in the choices for its enhancement or to give one's opinion on the use that is made of it – even though this convention does not refer explicitly to the concept of the common good. In this respect, INCULTUM's conception of the territorial tourism project is perfectly in line with the spirit of the Faro Convention action plan.

The rules for managing the commons are generally based on the know-how of the communities concerned, and on sharing principles that are neither those of public property nor those of private property. In her work on water, Sandrine Simon (2021) explains that the rediscovery of traditional know-how and community-based management methods for the commons, which are "intangible" components of the commons in their own right, "enriches and concretises the practice of the commons insofar as it emphasises the centrality of the actors concerned, and conveys the experiences, modes of governance and values that help to build the collective associated with it". In the spirit of the Faro Convention, heritage communities are often seen as the "behind-the-scenes driving forces [that] are autonomously formed groups that act independently to create their own narratives, claim sites and practices, and take responsibility for a particular heritage" (Joannette and Mace, 2019), but the point of view defended here is that, as soon as the political conditions allow, it is highly beneficial to establish close consultation with the public sphere, through appropriate governance, to sustainably manage the commons.

We speak of a territorial common when the common is circumscribed within a defined territory, whose uniqueness it helps to affirm. In this case, the associated heritage community will also be referred to as a territorial heritage community. This community may include permanent residents, secondary residents, occasional visitors and often also members of the diaspora, i.e. people who have emigrated from the territory in question and settled in other places; its members may have a variety of skills and social, economic and political roles, which is what makes it so rich.

The question of the territorial commons has been explored in depth by the Italian Territorialist School. It defines attachment to the commons as "an awareness, acquired through a process of cultural transformation by the inhabitants, of the heritage value of the territorial commons, both material and relational, as essential elements for the reproduction of individual and collective, biological and cultural life" (Magnaghi, 2010, p. 108).

This awareness and the transition from the individual to the collective are conceived through the collective recognition of elements of community in open, relational and supportive terms: "representing the different specialised knowledge of territorial heritage is the first step towards knowing, managing and socially reproducing territorial heritage" (Poli, 2018, p. 117). Beyond this, it is a question of relearning how to live in places by drawing on their specific features, as advocates of bioregionalism have been claiming since the 1970s (Berg and Dasman, 2019), a concept that is making a strong comeback as climate change accelerates. From another point of view, the aim today is to restore a 'landscape thinking' that was erased throughout the 20th century by the standardisation of spatial planning practices (Berque, 2015).

## 2.3    INCULTUM pilot projects: a range of territorial scenarios

Eight out of ten pilots concern rural areas that have become marginalised over the course of the 20th century from the point of view of economic activity, resulting notably in a significant demographic decline (80% for no. 6).

In all cases, this marginalisation is correlated with a mountainous geography (or insular, for pilot no. 10) that has kept these areas away from both automobile communication routes and the technical revolution in agriculture. The latter experienced accelerated development from the mid-20th century, resulting in the industrialisation of production systems. This industrialisation found more favourable conditions in the plains due to the absence of topographical constraints that could impede its implementation.

The result of these shared geographical and historical conditions today is a socio-economic situation that has many points in common: low population density, an economy that is not very dynamic, particularly in the services sector, and, conversely, landscapes that have been preserved in the sense that they have escaped the trivialisation associated elsewhere with the economic development of the 20th century (standardised housing and farming practices, omnipresent road infrastructure, etc.). The fact that the technical revolution in agriculture has not been completed in these areas has also resulted, to varying degrees, in a form of resilience linked to the preservation of traditional know-how.

The areas where the pilots are based are also marginal from the point of view of tourist attractiveness, in the sense that they do not have an emblematic point of attraction (monument, grandiose landscape). Nevertheless, this has spared them the consequences of tourist overexploitation, which many areas in the European continent are currently experiencing. In any case, the presence of highly active tourist attraction poles in close proximity (a coastline developed for beach tourism, a museum city, etc.) has not been sufficient to spontaneously trigger the development of tourism in these territories.

Two pilots do not fit this pattern. Pilot no. 2 (Campina de Faro, Portugal) concerns a densely populated coastal plain that has remained untouched by seaside tourism. Its uniqueness takes the form of the omnipresent material remains of an age-old agrarian system based on irrigation, which was also undermined by the new technical processes developed in the second half of the 20th century. Pilot no. 9 (Historic Graves of Ireland) focuses on a particular heritage motif, ancient cemeteries (particularly those associated with the great famine of the mid-19th century), on a country-wide scale, but it should be noted that the inventory focuses in most cases on sites located in rural areas that present the same characteristics as those studied by the other pilots.

Each pilot is also characterised by the emphasis placed on heritage features likely to serve as a basis for the tourist appeal of the area concerned. In most of these cases, the aim is to highlight a feature of the area, physically marked in space and more or less closely associated with traditional skills or practices that are now being undermined by modernity, or even threatened by the disappearance of human memory. It is probably no coincidence that most of these motifs take the form of networks, whether they be paths of local interest (no. 6: Bibracte - Morvan, F), transhumance paths (no. 7 and 8: Aoos/Vjosa river valley, AL/GR) or of trans-regional interest (no. 5: Tuscan-Emilian Apennines, I), or irrigation systems (no. 1: Altiplano de Granada, ES; no. 2: Campina de Faro, P; no. 4: Monti di Trapani, I). These networks of routes for people, animals and water were assets used and managed collectively by rural communities, usually according to rules that were not codified but were the result of usage. In this respect, they were real commons, the reactivation of which is conducive to the development of collective projects, as well as being ideal routes for exploring the area on foot or by bike. The pilots concerned are therefore placing the development of tourism on these routes at the heart of their action.

For the pilots where the heritage theme is not a network (mining heritage for no. 3: Central Slovakia; ancient cemeteries for no. 9: Ireland), the design of discovery itineraries nevertheless remains central to the projects, since it is the organisation and, one might say, the staging of visitor movements that are considered crucial for the development of tourism in the areas, according to a strategy quite different from that of "hot spots", which consists of keeping tourists in an emblematic place as long as possible in order to increase their consumption there.

This chapter therefore looks in particular at the ways in which the networks linking the members of rural communities are being enhanced, and how discovery itineraries are being set up as levers in the project to enhance the area.

From the perspective of the human resources and collaboration implemented in the pilots, another shared characteristic should be noted – the significant involvement of the academic world in the projects, often taking a leading role. This can be attributed to the funding modalities, which are channelled through a targeted call for proposals specifically addressing this category of stakeholders. This characteristic can be seen as a bias in relation to the usual conditions under which regional tourism projects are developed, but this bias is conducive to innovation insofar as it makes it possible to mobilise skills that are not generally present in the field, usually through a strong multidisciplinary approach. The downside, however, is the obvious fragility of the experimental projects, which run the risk of collapsing as soon as the scientist-animators have left the field, when the European public funding comes to an end. The solidity of the governance put in place, thanks to the mobilisation of local players, is therefore a criterion for assessing the performance of projects that is at least as important as the ingenuity of the systems deployed. The innovation brought about by the pilots must therefore be assessed from two angles: technical innovation and innovation in use, if by innovation in use we mean the establishment of an organisation that ensures the initiative's sustainability.

## 2.4    Stakeholders and their involvement

The INCULTUM consortium shared the vision of the "quintuple helix" model proposed to describe desirable innovation at a time of socio-ecological transition, with a view to fostering "the formation of a win-win situation between ecology, knowledge and innovation, creating synergies between economy, society, and democracy" (Carayannis et al., 2012). This model envisages five sub-systems that need to be considered in a concerted manner to ensure the future of a social organisation: the political system, the education system, the economic system, civil society and the natural environment. The way to coordinate the "sustainable" evolution of the five systems is through the circulation of knowledge, seen from an interdisciplinary perspective, where knowledge brought to a sub-system has the ability to stimulate that sub-system to produce and inject new knowledge into the system.

This model therefore articulates three approaches: (1) a mapping of the stakeholders in the system under consideration, (2) the need for an exchange of knowledge between the stakeholders and (3) the iterative nature of the system, based on the permanent circulation of knowledge from one system to another. Its transposition to a regional or local scale can clearly inspire management systems for rural areas.

In the situations studied by INCULTUM, stakeholders can be mapped by classifying them according to two criteria: the geographical scale of their field of intervention and their nature (see Figure 2.1). We distinguish three geographical scales: the local scale, which is also that of the territorial community under consideration; the regional (or intermediate) scale, which is that at which tourist destinations are usually managed; and finally the upper scale, at which policies on tourism, territorial development and the protection of heritage (cultural and natural) are decided and controlled, this being also the scale at which research policies and media that may have an impact on tourist flows are organised. As for the nature of the stakeholders, we have distinguished three categories: civil society, the economic

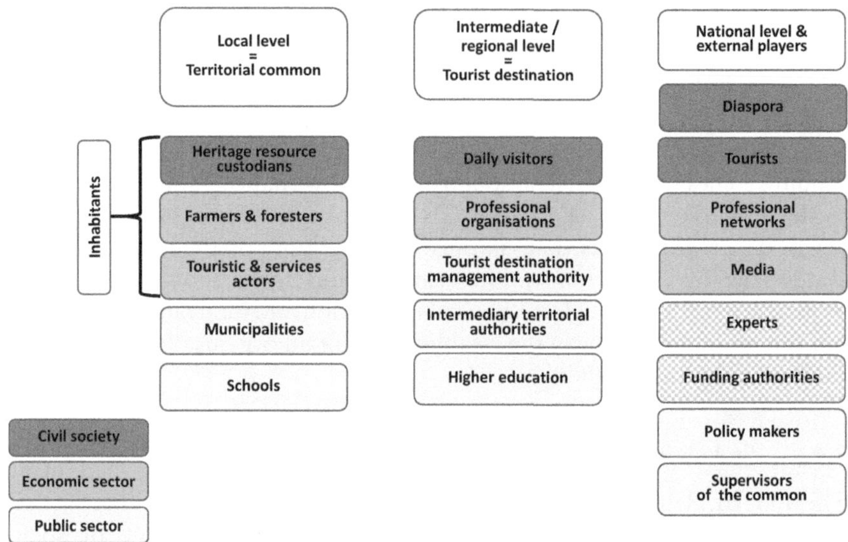


*Figure 2.1* Mapping the stakeholders in a common territory. Copyright: Bibracte (Flore Coppin et Vincent Guichard).

sector and the public sector. The public sector brings together stakeholders with a variety of profiles: political decision-makers, managers of sectoral policies such as tourism, heritage management authorities, education and research players. The economic sector largely overlaps with that of civil society, since, for example, a local player in the tourist industry is also a local resident who may be involved to varying degrees in social activities (often in associations).

Civil society operates on three geographical scales, with permanent residents on the local scale, day trippers on the intermediate scale, and tourists and members of the diaspora on the upper scale. The category of residents is the most difficult to define: ideally, they should all consider themselves custodians of the territorial commons, but in practice, a large part of the population is difficult to involve, including the "invisible" members of the territory, among whom are all the "little hands" in the service sector, without whom it would be impossible to organise a tourism offer. An important criterion for assessing the performance of a regional project is therefore its ability to involve the entire population.

## 2.5    The sustainable tourism wheel

The conceptual model developed as part of INCULTUM (see Figure 2.2) meets the iterative requirement expressed by the innovation helix models: one turn of the wheel strengthens the involvement, skills and contribution

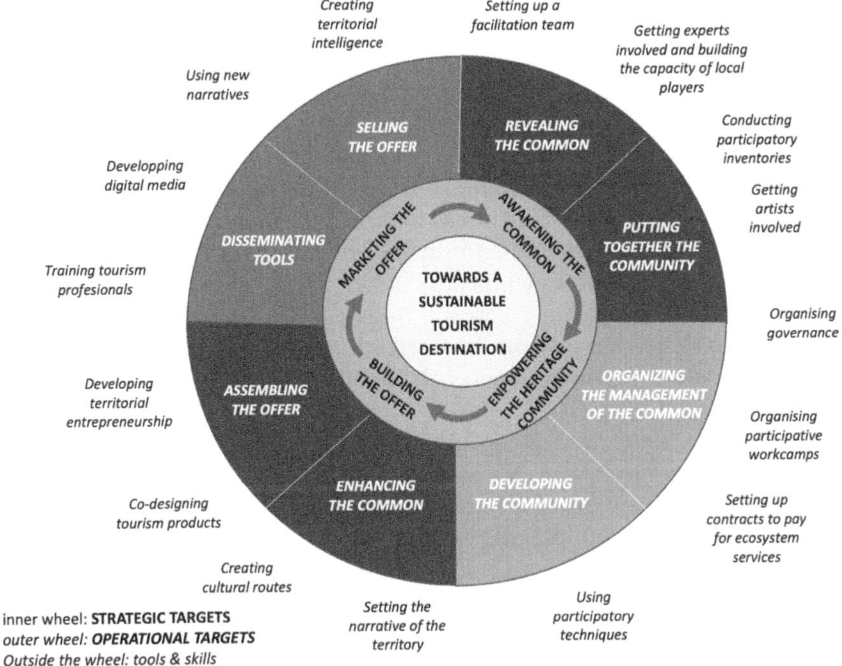


*Figure 2.2* INCULTUM's sustainable tourism wheel. Copyright: Bibracte (Flore Coppin and Vincent Guichard).

of the stakeholders, which, in a virtuous process, enables the next turn to be approached with more favourable initial conditions.

Each turn of the wheel involves a logical sequence of four stages, each of which responds to a specific strategic objective:

- Stage 1: reveal the common ground and build the community
- Stage 2: organise the management of the common ground and develop the community
- Stage 3: building the tourism offer
- Stage 4: marketing the tourism offer

Each stage can itself be divided into two operational sub-objectives. The pilot projects were committed to testing actions that contribute to these sub-objectives. These actions are listed at the periphery of the wheel in the logical order in which they are to be implemented, together with the tools and skills that underpin them. These actions have a more or less marked experimental character, and it is rather through their concerted and coordinated mobilisation that an innovative approach to the management of a tourist destination is expressed.

## 2.6   Step 1: unveiling the territorial commons and forming the community

The pilot projects of INCULTUM share a common foundation: the identification of the constituent elements of each lived territory that evoke shared attachment among its inhabitants. Far from being limited to elements of the "grand heritage" with national protection, it turns out that multiple elements of our daily environment contribute to the attachment we have to our place of residence and, consequently, to the well-being we feel there.

Often, this object of attachment is linked to the natural resources of the territory: water, forests, or to a rural heritage, whether tangible or intangible, closely tied to these resources – traditional agricultural systems, mines, transhumance or pilgrimage routes, depopulated ancient villages, traditional craftsmanship, and more.

In all cases, these heritage resources have been subject to abusive extraction (forests, mining resources), unwise use (water), or even to forgetfulness or memorial neglect (abandoned cemeteries and graves, transhumance routes used by minorities, depopulated villages).

However, these resources generate various types of ecosystem services and contribute to what some landscape-oriented development actors refer to as "territories of well-being": protection and generation of fertile soils, aquifer recharge, ecological corridors, carbon sinks, pride and self-esteem of local communities, and more.

These resources can be considered as commons to be reclaimed, especially when their management has been neglected in the recent past. The pilot projects aimed to initiate concrete actions in their favour: how to inventory and qualify their uses? What actions can be implemented to bring them out of oblivion by reactivating people's memories and identifying the traditional know-how necessary for their maintenance? How to quickly initiate restoration work, even modest ones, to start a dynamic process? And above all, how to promote the reappropriation of these resources?

### 2.6.1   *Setting up a project management team*

INCULTUM's experience has shown that the project management team plays a fundamental role at every stage of the process. The pilot projects have put in place project leaders who are positioned as coaches, responsible for the emergence, leadership and structuring of heritage communities. This role involves acting as referee, analysing complex situations and mobilising the expertise needed to overcome stumbling blocks, persuading both the local community and the external bodies on which the success of the approach depends, starting with the authorities in their role as guarantor of compliance with the law and guardian of public funding, and last but not least, empathy with the local project and its stakeholders. To all these conditions must be added the need for long-term investment on the ground, which is undoubtedly the greatest challenge for this type of project, because the funding required is very difficult to mobilise, especially as the current trend, at both European and national level, is to favour the mobilisation of expertise from outside the area in the form of short-term missions rather than its long-term implementation. This approach, which is now widespread, does not make it possible to capitalise on the expertise in the regions or, in other words, to create the conditions for a sustainable movement on the sustainable tourism wheel.

### 2.6.2 *Mobilising experts and building the capacity of local players*

When it comes to setting up holistic regional projects, of which tourism is only one component, it is essential to mobilise multidisciplinary expertise over the long term and to create the conditions for strengthening expertise that is firmly rooted in the region.

On the first point, all the pilot projects have involved university teams in the front line, as has already been mentioned. The challenge is therefore to ensure that the expertise mobilised does not disappear when European public funding comes to an end. One solution, which is being implemented to a greater or lesser extent by the various pilot projects, is to set up project areas as laboratories for territorial experimentation, able to host teams of scientists over the long term for research-action projects that capitalise on their results over the years, making these areas showcases for good practice and privileged spaces for consultation, able to attract public support over the long term.

On the second point, we need to put in place action-training systems that give learners an active role, enabling them to participate in the development of questions and methods adapted to their local context, in the spirit of "Science with and for Society". While it is not possible to train the whole population, it is more realistic to identify resource people in the local ecosystem and to rely on their ability to pass on knowledge. Making the most of the expertise of these people is a key factor in the success of this type of project.

### 2.6.3 *Conducting participatory inventories*

One of the priority tasks of the INCULTUM pilots was to mobilise and connect people who are attached to their common heritage. To do this, conducting participatory heritage inventories proved to be an excellent way of revealing shared reasons for attachment and initiating the establishment of a heritage community.

Typically, these inventories were conducted in the form of days during which a collective – often a municipality – was asked to show the group the places it considered most interesting in its living environment. In the case of the French pilot (no. 6), it turned out that the field trip organised by the village community benefited from being supplemented by a collegial examination of the oldest cadastral maps, dating from the first half of the 19th century and an original copy of which is still kept in the town halls (something that many residents are unaware of). These very precise maps, drawn up at a time when the rural population was at its peak, are an excellent tool for deepening discussions within the group and raising awareness of the importance of the heritage represented by the landscape. The effectiveness of these inventories has always been enhanced by the presence of experts (historians, archaeologists, geographers, etc.) in the group and by the use of technological tools to help with reading the landscape (such as LIDAR surveys) or locating and recording heritage features (such as overflights by drone or the use of a web GIS on a tablet). For example, in the case of the Spanish pilot (no. 1), the digitisation of the traditional Castril irrigation system and the use of the QField webGIS to locate points of interest in the area contributed to scientific knowledge of the hydraulic heritage and the drafting of a discovery route. In Ireland (no. 9), the identification of burial sites using drones, GIS mapping and photogrammetric modelling produced the planimetric documents needed to identify conservation tasks in consultation with professionals and local communities. In Albania (no. 8), following the heritage inventories, local meetings provided a forum for discussion with members of the Vlach community. They helped to reactivate the memory and open a debate on the future of Vlach culture, which led to local institutions taking this heritage into account. In Ireland (no. 9), a digital platform has been created to improve the quality of volunteers' involvement with local heritage communities. "Passers-on" can find the tools they need to pass on the components of the heritage commons and the ways in which it is managed, while the work done by local communities is promoted.

### 2.6.4  *Involving artists*

In a rural context with a sociologically fragmented population, artistic mediation is a relevant tool for establishing dialogue, renewing the perspective of residents on their living environment, and assisting in a shared narrative of the territorial project (see below). In France (no. 6), a member of the animation team dedicated efforts to mobilising artists and implementing interventions, following the concept of "Political Arts" developed by the sociologist and philosopher Bruno Latour (2021, 2023). This work aims to showcase the natural and cultural heritage of the territory by envisioning it as an artistic stage, a multifaceted agora where different audiences come together. Artists in residence propose encounters, listen to the residents and embrace memories and stories. Others, who live in the area, are engaged to offer workshops for children and the elderly.

## 2.7  Stage 2: organising communal management and developing the community

### 2.7.1  *Organising governance*

If the community project is to be sustainable, a system of governance must be put in place and spelled out, giving each stakeholder an appropriate place, according to the adage "a place for each stakeholder and each stakeholder in his or her place". The governance plan is in fact the expression of the rules that the heritage community adopts. Putting it in place is a long-term and largely empirical process. It is in fact necessary to create a shared vision, not only of what constitutes a common heritage, but also of the legitimate ways in which it is used, the consequences of the actions of some on the uses of others, and the perimeter of the people authorised to claim access to these uses. And as it is not enough to establish rules "once and for all", but to bring them to life as successive generations and new visitors or residents enter the area, it is important to establish permanent forums for dialogue to ensure that these standards evolve and, above all, that the desire of the members of the community to work together is maintained.

The diversity of local situations makes it impossible to propose a standard model that can be adapted to all circumstances, unless it is reduced to an over-simplistic scheme: in all cases, there must be a plenary assembly of stakeholders, a steering committee and an operational mechanism.

In the governance arrangements for pilot no. 6, the positioning of the operational team at the interface between the world of research and the other stakeholders makes it possible to meet the requirement for the circulation of knowledge set out in the five-blade propeller model. In this task, the team is mobilising a particular category of stakeholders: players from the worlds of culture and education, as well as artists.

Generally speaking, the collectives mobilised by the pilot projects are not necessarily recognised organisations or organisations with official status. More or less structured depending on the case, they bring together people who have decided to take on the interpretation and management of a particular form of their heritage, sometimes of their own accord or as a result of the impetus provided by the pilot projects.

For example, at Bibracte (no. 6), the activation of the community is based primarily on the Rural Paths working group. The group is led by a member of the operational team mentioned above – in this case, the INCULTUM project manager for the start-up phase of the action; it meets on a regular basis and is involved in a number of tasks: taking stock of the network of paths and its uses, drawing up a shared management plan, identifying discovery routes to be developed, promoting tourism. This group has proved to be an effective vehicle for strengthening the community, thanks to the widely shared interest in the paths and, in so doing, the diversity of its members' profiles and the interest shown in it by the local councils.

### 2.7.2   *Organising participative workcamps*

To get local residents involved over time, it is important to anchor the project in history and over time, in particular through annual events and rituals such as participatory heritage maintenance projects (France, Spain) or long-term educational initiatives in schools to raise the awareness of different generations (Slovakia).

### 2.7.3    *Setting up contracts to pay for ecosystem services*

As part of the Spanish project (no. 1), the irrigation communities made a commitment to the municipalities to maintain the traditional irrigation systems and the new associated cultural itineraries within the framework of agreements. It is interesting to note that these agreements are favourably received by the communities, even though they do not provide for payment in cash, but only for support in kind from the municipalities, such as the loan of equipment or the provision of labour. It seems that the decisive factor is more of a symbolic nature: through such agreements, the local authority and the residents it represents recognise the know-how and usefulness of the work of the members of the irrigation communities.

### 2.7.4    *Using participatory techniques*

In France (no. 6) and Italy (no. 5), artistic residencies have been organised in collaboration with the heritage community to create links between local people, gather their perceptions of the evolution of the area's landscapes and collect local stories. In Italy, these residencies led to the creation of a play and in France to the publication of a collection of illustrated booklets, which are a poetic and sensitive way of strengthening the attachment of local people to the area, attracting new visitors and stimulating dialogue and training opportunities for local players.

## 2.8    Stage 3: building the tourism offer

### 2.8.1    *Telling the story of the region*

Storytelling is a classic – and effective – tool in tourism marketing, when it is used to encourage people to discover a destination by highlighting, often in a somewhat insincere way, the uniqueness and "authenticity" of its heritage and landscapes, or the quality of the welcome offered by its inhabitants. We take a different approach here. The target audience for the stories is first and foremost the area's inhabitants, with the aim of sharing with them a positive vision of their region, based on themes and values linked to the management of the community and the way the area is inhabited, past and present (Scheyder et al., 2022).

It is only when such stories are shared within the community that it becomes legitimate to share them with visitors. It is not only a way of resonating with the area, but also of renewing collective representations and imaginations. By conveying to visitors their attachment to their territory, heritage communities encourage them to consider the relevance of historically sustainable skills and lifestyles. They raise visitors' awareness as "temporary residents" of the threats posed by climate change and inspire them to take action to care for and protect heritage resources. "Understanding the changing nature of heritage can help mitigate the general fear of change and loss of rights that sometimes arise from conflicting local-global narratives" (Horizon 2020, 2023). What's more,

> Stories about bioregions or the territorial commons [...] can also nurture an imagination of the right balance of needs, balanced resource management and co-existence with other living things. Finally, alliances of players can nurture an imagination linked to symbioses, systems and ecosystems, networks and territorial solidarity; and neighbourhoods (districts, urban or rural villages, collaborative housing, street squares, shared gardens) an imagination linked to proximity, short supply chains and mutual aid. In this way, these performative narratives contribute to the emergence of a new collective imagination that involves and unites citizens around the challenges of transition.
>
> (Scheyder et al., 2022)

In this perspective, discovery routes serve as an effective narrative framework for storytelling about the territory, as one progresses through walking. By encouraging encounters with members of the local community, these routes add depth to the territories. The paths taken (re)claim their traditional role as "third places" and spaces of sociability where both visitors and residents can geographically anchor themselves through the emotions and affective feelings experienced during their journey. This allows for the cultivation of a "presence ecology" that resocialises and rehumanises us, through performative narratives that convey an imaginary and provoke a desire to take action.

For instance, in Portugal (no. 2), narratives collected through interviews with former farmers, considered as guardians of memory, serve as the foundation for the story map of new routes that evoke the natural and cultural history of the coastal plain of Algarve, the traditions of solidarity, frugality, and adaptation within its farming community and more.

An article published in The New York Times (Méheut, 2023) about the Spanish project (no. 1) is an excellent example of media receptivity to these new narratives, as evidenced by its title: "Facing a Future of Drought, Spain Turns to Medieval Solutions and Ancient Wisdom".

### 2.8.2   *Creating discovery routes*

Most of the pilot projects have set themselves the operational objective of creating new cultural tourism routes to discover local heritage resources. Three years after the initiative was launched, some are well on the way to being homologated by official bodies.

The creation of these routes has proved to be an effective way of bringing stakeholders together, thanks to the diversity of the people and organisations involved: local councils, tourism stakeholders and users of the paths in all their diversity (farmers, foresters, hunters, sports enthusiasts, walkers). In this specific case, the tourism project is a lever for designing, maintaining and developing new discovery routes, which strengthens the heritage community, provided that the approach is carried out in partnership.

### 2.8.3   *Co-designing tourism products*

Enrico Bertacchini (2021) sees in the concept of the commons the possibility of enriching the economic analysis of cultural heritage and overcoming the problems posed by its conservation and management. To this end, he proposes an integrated approach to defining the role of heritage in sustainable development, based on three key concepts that combine and interact:

– Cultural capital – heritage is a fixed asset that produces other goods and services, so the cultural values it embodies are put on the same footing as the economic values it can generate;
– The cultural district, which emphasises the entrepreneurial, organisational and territorial dimension of the production of goods and services linked to heritage assets;
– The cultural commons, which opens the door to the consideration of local communities and questions the governance of the various dimensions of heritage as a shared resource.

Collaborative economy is a preferred avenue to successfully carry out this project. It brings together economic activities based on the sharing or pooling of goods, knowledge, services or spaces, and on use rather than possession. It is based on networks or communities of users and aims to pool resources, both tangible and intangible. It is based on horizontal exchanges and generally relies on digital platforms as intermediaries between users.

> The various theories of the commons have notably contributed to the emergence of the collaborative economy by formalising the notion of peer-to-peer exchange and community projects, but also by renewing the approach to governance within collective projects ... by drawing inspiration from the principles of social utility, democratic governance and controlled profitability of the social and solidarity economy, [the collaborative economy] is a major lever for transforming society, particularly in terms of the ecological transition and the development of social ties and territorial dynamics.
>
> (Avise, 2020)

Some pilots have thus devised economic systems based on the collaborative economy to manage the heritage commons: peer-to-peer exchange via "payment for services" agreements (no. 1) (see above), participatory finance via crowdsourcing (no. 9), voluntary work via heritage restoration sites (no. 1 and 6), contributory production of open source 3D technologies, free collaborative platform (no. 9), etc. On a different theme, the Greek pilot (no. 7) uses open source technology to deliver a solution for monitoring the sensitive ecosystem of a lake located on the Discovery Route.

In Ireland (no. 9), the *Historic Graves* initiative began as a collaborative project based on crowdsourcing. Over time, this initiative has revealed its strong tourism potential, as demonstrated by the many comments the platform constantly receives from users from all over the world in the diaspora who are planning to visit Ireland because they have found the exact location and state of preservation of a family member's grave.

For some projects, volunteers are key players in the management of the resource. The volunteer work camps organised by some of the pilots for the restoration and enhancement of certain elements of their heritage are an opportunity for young volunteers to participate in the enhancement of rural heritage and to be made aware of the issues involved in the ecological transition of territories in the context of climate change.

These initiatives maintain control and autonomy over the management of the commons, and the benefits accrue directly and in a circular fashion to the heritage communities, generating positive impacts. They involve a participatory approach, giving the heritage community a leading role. The collaboration between the scientists and experts coordinating the project and the local communities also serves to encourage and empower them.

### 2.8.4   *Developing territorial entrepreneurship*

The new cultural discovery routes are federative initiatives for the local economy. The pilots have modelled an integrated approach to the tourism economy, using the concept of "tourism as a tool" to encourage the development of territorial entrepreneurship and create a local dynamic with a positive social and environmental impact through cooperation between stakeholders.

Territorial entrepreneurship is defined as "an entrepreneurial movement that reinvents new, more collective ways of doing business, with the aim of generating responses in favour of a more rooted, sustainable and inclusive economic development" (Baudet, 2017, p. 72). It can cover a wide range of areas: safeguarding traditional activities, maintaining the rural socio-economic fabric, promoting local products, developing new activities, etc. It is an alternative to public action and private entrepreneurial projects. It can take the form of a traditional business or a legal form derived from the social economy.

Tourism is therefore a real lever for sustainable economic development in the region, and the creation of tourist offers and services should be seen as a resource for the tourist destination, entrepreneurs and local residents. Encouraging tourism is an integral part of INCULTUM's territorial initiatives.

The aim is to inject the added value of the collective entrepreneurial project into the region, to fuel circular and social-economic flows.

In INCULTUM project, farming communities and local producers are key players in the management of resources, the upkeep of the agrarian landscape and a basis for the development of sustainable tourism, as they contribute to the development of farm-to-fork circuits and agro-tourism (with farm visits, for example).

In Portugal (no. 2), the involvement of the farming community in agro-ecology and tourism has revealed a "technological unit" (traditional hydraulic infrastructure) and a "social unit" (local community, farmers' association) through the study, rehabilitation and dissemination of its importance and added value for society and the environment.

Similarly, in France (no. 6), a collective of local farmers has been set up to work together to take over farms, create links with the service economy, particularly tourism, and strengthen solidarity within farming communities. The collective is organised as an association and is recognised as an Economic and Environmental Interest Group (GIEE), a label awarded by the French Ministry of Agriculture, which gives visibility to its actions.

The communities that organise themselves on the scale of a territorial common can thus be seen as cooperative territorial "enterprises" in which all the players contribute to shared objectives with a view to a "return to the territory as a common good" (Magnaghi, 2014). Here, the territory is seen as a system of relationships and actors defined not by administrative and political boundaries, but by the density of the relationships that develop within it.

From this perspective, the commons management model goes further than the participatory mechanisms derived from the delegative model of representative democracy, by including a stage of co-decision with citizens through bottom-up processes of social innovation. The aim is to co-organise and engage in dialogue with communities that are not content with the delegative space, but act.

The role of the public authorities is to detect these initiatives, support them and help them to spread. In this configuration, the local councillor takes on the role of catalyst, provided that he or she has a culture of supporting bottom-up and social innovations. The heritage community, for its part, becomes a place where social innovation can be identified and public policies inspired.

Support for entrepreneurial social innovation encourages the creation of new projects and new economic activities based on local development issues and local players and resources. An important aspect of innovation is the hybridisation of economic sectors, which today suffer from having been specialised since the middle of the 20th century as part of the establishment of the Common Agricultural Policy, resulting in increased vulnerability to the vagaries of the market and the climate. This hybridisation amounts to a return to an old situation where "you don't put all your eggs in one basket" by having one foot in agriculture and another in another activity (forestry, building trades, industry, etc.). In the context of INCULTUM, the priority is to enable farmers (and foresters), who are the "workers of the landscape", to enjoy a decent standard of living over the long term. To achieve this, the services and visitor reception sector offer a whole range of agritourism possibilities: food production to be sold in a short circuit on the local market, reception on the farm, development of a guiding or concierge activity.

## 2.9    Stage 4: marketing the tourism offering

### 2.9.1    *Training tourism professionals*

Considering tourism as a tool for integrated heritage management and as a component of rural territorial strategies means that we need to think about adapting the academic courses that lead to professional qualifications.

Considering shared heritage as an area for local development, integrating the tourism, agricultural, craft, scientific and educational dimensions, means strengthening the skills of those involved in the subjects of ecological and social transition and rural development policies.

As mediators in direct contact with the public, tourist guides play an important role in raising visitor awareness of the principles of sustainable tourism, heritage preservation, the environment and biodiversity. The training of guides, including their acquisition of expertise in the specific characteristics of the areas in which they work, is therefore a cornerstone in the creation of new tourist offers for the pilots.

### 2.9.2    *Developing digital media*

The new digital commons are co-produced and open citizen resources that are recognised for their sobriety, resilience and collaborative virtues. As part of their actions, some pilot projects (no. 1, 6, 9) have experimented with participatory digital mapping tools for their heritage commons. These platforms are sometimes used as collaborative management and awareness-raising tools for territorial commons (Ireland, France).

However, they are costly to develop and are not optimised for destination marketing. To make their new tourist itineraries visible, designers are forced to use proprietary digital platforms in a highly fragmented market. The experiments carried out as part of INCULTUM have therefore convinced us of the benefits of developing tools that fully meet the criteria of the new digital commons, not only for sharing tourist itineraries more effectively, but also for inventorying and managing them by mobilising the user community, whose role could be to assess the relevance of the proposals.

### 2.9.3    *Using new narratives*

If a tourist destination corresponds to a territory experienced and managed as a common, then discovering it can become a new, empowering way of learning and being inspired by stories about the ecological transition, local resilience or heritage in all its meanings.

Some pilots have created digital storytelling platforms that are proving to be effective tools for promoting new tourist destinations (see pilot no. 3, for example). The narratives conveyed on these platforms, in which the heritage character of the place is expressed by local people, contrast with the dominant tourist imaginary of the "conventional" tourism industry, which is defined in terms of the attractiveness of the destination and the rhetoric of tourism promotion.

Finally, some pilots have chosen to develop their own regional brand (no. 2, no. 6), which brings together local players around a common project and shared values, in order to cooperate and raise their profile with different audiences. The aim is to promote activities with a strong social impact or existing local industries that are involved in local supply chains to the local population and tourists. These brands respond to the need for traceability of products or services sold, particularly in the food sector. They help to enhance local identities and to associate the terroir, heritage or local know-how with the quality of the products or services associated with them, as well as with economic dynamism, sustainable development and social, environmental and economic impact. They also help to generate a strong identity for the area, attracting local entrepreneurs, new residents and financial partners.

### 2.9.4   *Creating territorial intelligence*

The use of strategic intelligence tools is essential for getting to know visitors, analysing their behaviour and understanding the impact of tourism on heritage resources. What's more, the shared construction of an objective and informed diagnosis combining quantification and perception of tourism makes it possible to create territorial intelligence, particularly in the tourism economy sector.

In a development model for territorial projects where the creation of value is dissociated from the use of common resources, where economic models are based on positive impacts and not on volume and quantity, and where players rely on cooperation rather than competition, sharing this knowledge with stakeholders guarantees the sustainability of the territorial project.

In this way, in response to local mistrust of the tourism project, the French pilot (pilot no. 6) used a mechanism designed to create a form of territorial tourism intelligence. Through a local tourism observatory, the aim is to overcome preconceived ideas and objectify knowledge of tourist activity by means of in-depth surveys produced and analysed with visitors, local residents, economic players and local decision-makers. By repeating the survey over the years, it is possible to monitor changes in the behaviour of both visitors and local stakeholders and to redirect the area's tourism strategy.

The Swedish pilot project (no. 10) has developed a system for understanding visitors and their spatio-temporal behaviour, using a mobile phone data collection system to provide information about visitors and GPS loggers to geolocate visitors' positions. Netnography completes this system by examining visitor discourse on social media. This system provides decision-makers and players in the tourism industry with concrete, up-to-date data on visitors' perceptions of the region, to inspire and guide tourism strategies.

## 2.10 Conclusion: tourism as a factor in the cohesion and resilience of rural areas

The uncontrolled development of tourism in rural areas has undesirable effects that are well known and threaten the European area more than ever at a time when it is becoming increasingly attractive in the context of growing insecurity on a global scale and the need for sobriety if we want to keep the planet liveable, which in the field of tourism means choosing destinations that are less remote and explored more frugally.

This is the context in which INCULTUM's thinking and experimentation were developed, in the conviction that, properly managed, the tourism sector can have a positive impact on the areas concerned and their inhabitants. In our case, the areas concerned were marginal – in terms of their demographic and economic decline – and rural, which in return meant that they could maintain tangible and intangible heritage features that had disappeared in more active areas, and that tourism currently played a very small part in their economy.

It so happens that the pilot projects have all made the choice – initially uncoordinated – to create new visitor itineraries, making them the preferred means of discovering the heritage resources of the territories. These itineraries are based on communication or water distribution networks that were once managed as common property, and which are proving to continue to arouse a strong attachment among the populations concerned.

All the pilot projects have therefore based their action on the attachment of local communities to elements of the heritage of their territory. These heritage elements (tangible or intangible) can be described as territorial commons. The involvement of associated heritage communities is an essential lever for the development of sustainable tourism initiatives. In turn, the hypothesis defended by INCULTUM is that building a tourism offer based on such commons can strengthen the heritage community and, in so doing, help to preserve the commons and community cohesion.

Putting such an offer into tourism requires a suitable narrative. It can usefully mobilise the region's human resources to support visitors, as well as a range of digital tools – which are still far from perfected – to inventory, manage and share heritage resources, or analyse visitor behaviour.

If a tourist destination can be identified with a territorial common that is experienced as such by its inhabitants, the trip can become an inspiring experience, made up of genuine encounters and a concrete confrontation with the challenges of the ecological transition.

By considering tourism as a tool at the service of territories, INCULTUM has modelled an integrated approach to the tourism economy with a view to developing a form of territorial entrepreneurship that encompasses the different sectors of activity and promotes their hybridisation, by ensuring that as many players as possible are pluriactive and involved in both the primary sector of the economy (agriculture) and the secondary and tertiary sectors, in contrast to the organisation into compartmentalised sectors that was promoted throughout the 20th century.

The use of strategic intelligence tools is essential for creating knowledge about visitors and understanding the impact of tourism on heritage resources. Sharing this knowledge with stakeholders creates a "territorial tourism intelligence" that guarantees the sustainability of the territorial project. More broadly, the success of the regional project requires the long-term implementation of a suitable coordination and governance system, which is undoubtedly the most difficult condition to meet at a time when public action is seeing its resources reduced and more often than not limited to short-term expertise missions that do not allow this expertise to be capitalised on in the regions where it is essential. In this context, the best we can hope for is that local entrepreneurship will be effective enough to generate the subsidies needed to pay for expertise.

Taking a step back, the tourism approach promoted by INCULTUM aligns with a

> Self-governance culture: mindful of the territory, which does not entrust the sustainability of development to machines or external decision-making centers. Instead, in a world inhabited by a multiplicity of development styles, it relies on a rediscovered environmental wisdom and on inhabitants capable of once again producing territory.
>
> (Magnaghi, 2010, p. 36)

The challenge now is to create conditions for this "self-governance" to fully assume itself, particularly in terms of economic self-sustainability.

## Bibliography

Avise. (2020). *Dossier ESS & économie collaborative*. Paris: Avise. www.avise.org/dossiers

Baudet, S. (2017). *Accompagner l'essor d'un entrepreneuriat de territoires*. Paris: Territoires Conseils [available online].

Berg, P., & Dasman, R. (2019). Réhabiter la Californie. *EcoRev'*, 47, 73–84. https://doi.org/10.3917/ecorev.047.0073

Berque, A. (2015). *Thinking Through Landscape*. New York: Routledge.

Bertacchini, E. (2021). Capital culturel, district culturel et biens communs. Vers une approche économique intégrée du patrimoine et du développement durable. *Situ. Au regard des sciences sociales*, 2. https://doi.org/10.4000/insituarss.883

Carayannis, E.G., Barth, T.D., & Campbell, D.F. (2012). The Quintuple Helix innovation model: Global warming as a challenge and driver for innovation. *Journal of Innovation and Entrepreneurship*, 1, 2. https://doi.org/10.1186/2192-5372-1-2

Faro Convention. (2005). *Council of Europe Framework Convention on the Value of Cultural Heritage for Society*. Strasbourg: Council of Europe. https://rm.coe.int/1680083746

Horizon 2020 (2023). *Policy Recommendations by Horizon 2020 Sustainable Cultural Tourism Projects*. Bruxelles: European Commission [available Online].

Joannette, M., & Mace, J. (2019). *Les communautés patrimoniales*. Québec: Presses de l'Université du Québec.

Kunstler, J.H. (2005). *The Long Emergency: Surviving the Converging Catastrophes of the Twenty-first Century*. New York: Grove Atlantic.

Latour, B. (2021). Comment les arts peuvent-ils nous aider à réagir à la crise politique et climatique ? *L'Observatoire*, 57(1), 23–26. https://doi.org/10.3917/lobs.057.0023

Latour, B. (2023). *How to Inhabit the Earth: Interviews with Nicolas Truong*. Cambridge: Polity (transl. of Habiter la Terre, entretiens avec Nicolas Truong. Paris: Les Liens qui Libèrent, 2022).

Magnaghi, A. (2010). *Il progetto locale. Verso la coscienza di luogo*. Torino: Bollati Boringhieri (augmented edition).

Magnaghi, A. (2014). *La biorégion urbaine, petit traité sur le territoire bien commun*. Paris: Eterotopia.

Méheut, C. (2023). Facing a future of drought, Spain turns to medieval solutions and ancient wisdom. *New York Times*, 19 July 2023, https://www.nytimes.com/2023/07/19/world/europe/spain-drought-acequias.html

Ostrom, E. (1990). *Governing the Commons. The Evolution of Institutions for Collective Action*. Cambridge: Cambridge University Press. https://doi.org/10.1017/CBO9780511807763

Poli, D. (2018). *Formes et figures du projet local: la patrimonialisation contemporaine du territoire*. Paris: Eteropia.

Scheyder, P., Escach, N., & Gilbert, P. (2022). *Pour une écologie culturelle*. Paris: Le Pommier.

Simon, S. (2021). Réintroduire le patrimoine de l'eau parmi les communs. *Situ. Au regard des sciences sociales*, 2. https://doi.org/10.4000/insituarss.599

# 3 Place branding from scratch

## Naming, framing, and finding Campina de Faro

*Viktor Smith and Maximilian Block*

## 3.1 Introduction

### 3.1.1 Background, aims, and scope: challenges of the "below zero" scenario

Promoting cultural tourism in marginal (and any) places requires that at least some potential visitors know about the place and have some idea of what it is. This is what some scholars would call the existence of a *place brand* (Vuignier, 2017; Kaefer, 2021).

The concept of *brands*, whether relating to places or (more classically) companies, products, or services, has been subject to several, partially overlapping, definitions (cf. Maurya & Mishra, 2012; Oh et al., 2020). An essential common core shared by most of them is captured by ISO 20671–1:2021 (3.1) which defines a brand as an

> [...] intangible asset, including but not limited to, names, terms, signs, symbols, logos and designs, or a combination of these, intended to identify goods, services or entities, or a combination of these, creating distinctive images and associations in the minds of stakeholders [...].

Transposed to semiotic terms (e.g. Bussmann, Kazzazi, & Trauth, 2006: 397), a brand thus comprises an expression side (or "plane"), i.e. the outer cues by which it is recognized, including its name, and a content side (or "plane"), i.e. the images and associations that these evoke. In branding theory, the latter are often further divided into a *brand image*, i.e., the way the brand is conceived by its surroundings, and a *brand identity*, i.e., the way representatives of the branded entity itself conceive it and/or would like others to conceive it (cf. Kotler et al., 2016: 426). Aligning the two is a major concern in most place-branding efforts (cf. Stock, 2009; Rugaard, 2022: 30–31).

However, branding a place is in many ways different from branding a company, an organization, or a consumer good. The specifics include the absence of a single brand owner; the need to navigate across the often-conflicting objectives, interests, and actions of a plethora of stakeholders; and the complexity of what makes up a "place" in terms of geography,

DOI: 10.4324/9781003422952-3

culture, administrative-political organization, and so on (cf. Smith, 2015; Peighambari et al., 2016; Boisen et al., 2018; Kaefer 2021).

In addition to these challenges, which are all well recognized and much debated, the INCULTUM work has foregrounded certain additional complexities that seem to deserve more systematic attention in their own right (for public results, cf. INCULTUM, n.d.). This includes what some INCULTUM colleagues have so far informally, yet aptly, dubbed the "below zero" scenario found in many, though not all, marginal places targeted by the 10 INCULTUM pilots. What characterizes this scenario is (a) that visitors and locals alike are yet to recognize the place as something that tourists might consider visiting, (b) that reliable data indicating any baseline in that regard are scarce or absent, (c) that local stakeholders who might be interested in creating new tourism opportunities first need to be identified and brought together, and (d) that limited financial resources are available for any such initiatives.

The INCULTUM team at Copenhagen Business School (WP7) has dug somewhat deeper into the opportunities, obstacles, and dynamics of developing cultural tourism in precisely such a "below zero" context, taking the Portuguese pilot as an exemplar case. A first empirical inquiry to that end is reported in the following. It centres around an observation that may not be new as such, but which becomes particularly prominent in a "below zero" context: the need to have (and for people to know) a "name for it" as a first and crucial precondition for any more far-reaching branding endeavours. In this case: the name Campina de Faro.

The following key research questions are pursued:

*RQ1: What is the status and future potential of the name 'Campina de Faro' as a fixpoint for developing a more salient place-brand image of that area among potential visitors and elevating sustainable cultural tourism there?*

*RQ2: Which name-intrinsic and name-extrinsic parameters could be operated on to support this objective better in the future while keeping costs at a moderate level?*

*RQ3: Could any learnings from Campina de Faro be put to use also in territories covered by other INCULTUM pilots and could any initiatives taken in some of these be a source of inspiration in the case of Campina de Faro?*

However, first, we need to consider a few specifics of Campina de Faro as a potential target for cultural tourism. Having done that, we will introduce a conceptual framework suited for supporting a clearer understanding of the linguistic-semiotic, cognitive, and commercial variables in focus, including the specifics of naming and framing a place-brand on a budget. We then report a questionnaire survey conducted among visitors to the Faro-Albufeira area, supplemented by observations made by us during field research on the spot. On this background, we discuss the implications of the findings gained for RQ1 and RQ2 while also providing some tentative answers to RQ3.

### 3.1.2   *The case of Campina de Faro*

The Algarve Region is known for its sandy beaches, picturesque cliffs, and mild climate, stretching across the southernmost parts of Portugal and attracting millions of tourists annually (Luz, 2023). While cities like Faro and Albufeira are flooded with tourists, other places experience almost zero tourist visits. This disbalance can be immediately observed within the area known as Campina de Faro. Located in the hinterland of Faro between the historical cities of Faro, Olhão, and Loulé, Campina de Faro is a coastal plain rich in natural beauty and cultural heritage. It boasts a historical irrigation system dating back to Islamic times, flourishing gardens, orchards preserving traditional food production methods, and a distinctive local cuisine. It would thus appear to combine many attributes that could make it an attractive tourism destination, for instance, for daytrips made by sun-and-beach tourists at the Algarve coast. Yet, this is not reflected in the pre-INCULTUM situation.

Characteristically of the "below zero" scenario, exact data on tourism activity in this area are scarce. However, some tangible indications supporting the overall impression just described do exist. An exploratory study by Rugaard (2022) with a special focus on Danish visitors to Portugal (N=117) showed that among recent visitors to the Algarve (N=40), only five persons (12.5%) had ever heard of Campina de Faro and only three had been there. A survey conducted by the Portuguese pilot in which the majority of respondents were Algarve residents (88% of 134 respondents) showed that only 31.9% had heard about the place, and just 16.3% declared to have visited Campina de Faro (UAlg, 2023; Batista, 2023). Another indication, while not so far quantified in exact figures, is the limited online presence of the place which makes it challenging to find information on its exact boundaries, spots to visit, or even pictures.

The survey and field observations reported in the following add more nuances to this overall picture. As a background for the empirical work, we will first outline a conceptual framework suited for pinpointing the mechanisms through which a name can contribute (or not so) to shaping a brand image since this constitutes a pivotal element of the survey.

## 3.2   Conceptual framework

### 3.2.1   *Four perspectives on naming and framing*

At the heart of any branding effort lies the crucial task of naming and framing. A brand cannot be identified without having a name to connect it with and the exact conceptual content and associations that key audiences will come to connect with that name depend on how the name is framed by surrounding verbal and visual cues and people's real-life experiences with the branded entity itself.

Following Smith (2021), we here propose a four-layered analysis of the naming and framing processes involved in most goal-driven efforts to facilitate people's understanding of something by picking and shaping "the right words" for it, places included. The four perspectives summarized below are all well recognized in existing research, yet they are rarely viewed in integration. Many findings of mutual interest are scattered across multiple disciplines that only occasionally engage with each other: from linguistics, semiotics, and discourse analysis, through cognitive and perceptual psychology, to practice-oriented fields such as terminology management, public health promotion, marketing, branding, Public Relations (PR), and political communication; see also Entman's (1993) polemics on the scattered nature of framing research that largely remain valid today. More details on the underlying cross-disciplinary positioning and linkages are given in Smith (2021: 1–10, 77–79), yet here we will concentrate is on the resultant framework itself.

- *Perspective 1: Having a name for it.*

    "If you don't have a name for something, then as far as people are concerned, you don't have it at all", says an old wisdom here echoed in an ad for Tracey Communications (Smith, 2021: 11). As for places: It is impossible for potential visitors to become aware of the place's existence, seek more information about it, decide to go there, be delighted or disappointed, and to share their experiences with others, without having or learning a name for the place. Notably, the place may well have a name already (that can furthermore be harder to change than, say, when (re)naming a company or a product), but this is of no consequence to potential visitors if they do not know it.

    Moreover, for a place-brand image to emerge (whether consistent with the place identity experienced by the locals or not, cf. Kaefer, 2021; Rugaard, 2022; Kotler et al., 2016: 426), some degree of consensus is required across potential visitors' ideas of what the place is like. That is, there must be at least some common elements in the concepts that different individuals connect with the name. To begin with, these may well be sheer ad-hoc concepts (e.g., "it's probably some kind of historical monument") that people generate in an attempt to make some sense of the name. However, these may gradually develop into more elaborate, stable, and inter-subjectively valid ones (cf. Smith, 2021: 21–25; see also Barsalou, 1987, 2016; Murphy, 2010).

    Two sorts of influences jointly determine the final outcome of these processes, here summarized under Perspectives 2 and 3.

- *Perspective 2: The name's intrinsic communicative potential (The Joyce Principle).*

Some names are just names, while others (indeed, the most) give the recipient a certain hint, but nothing more, about what they *could* be referring to. In linguistic-semiotic terms, such names are referred to as non-arbitrary or motivated (Nöth, 1995: 240–256). In the marketing literature, the phenomenon is also known as the Joyce Principle (suggested by Collins, 1977) referring to the extensive use of suggestive (in that case non-existent) words like *smallfox* and *tattarrattat* in the literary work of James Joyce.

Following Smith's (2021: 28–29) further elaboration on Ullmann (1962), we will here distinguish between (a) *phonetic motivation* emerging from the way the name sounds (e.g., *crash boom, Wahooo!* (a waterpark in Bahrein)), (b) *graphic motivation* where the choice of fonts, colours, and surrounding imagery adds to the expressive potential (e.g., the brand name *Bake & Freeze* half in red and half in blue or the Spanish nation-brand logo *España* incorporating characteristic red, yellow, and black ornaments), (c) *semantic motivation* where a new meaning is assigned to an existing word through metaphoric or other semantic transfer (e.g. *house* (the music style), (computer) *mouse, The Big Apple* (for New York City)), and (d) *morphological (structural) motivation* which emerges when a name is composed of smaller units which jointly hint at the meaning of the whole name (e.g., *greenwashing, dark tourism, Cold Hawaii* (a surfing spot in northern Denmark) ... and *Campina de Faro*). More than one type of motivation may well be in play for the same name, for instance, emerging from different stages of its formation, e.g., *Baby Boomers* or, for *Campina de Faro*: because it is not only composite but also sounds in a way that suggests a certain geographical and lingua-cultural origin.

- *Perspective 3: The formation of the name's full meaning in running communication (The Juliet Principle).*

  While the Joyce Principe is crucial when a new name is launched or heard by somebody for the first time (*green tax* rings a different bell than *fuel tax*), its significance tends to fade over time. One may well use an *Apple* laptop or find a *monkey wrench* in one's toolbox without having any idea about why they are called so.

  On that background, some scholars and practitioners argue that the presence of any "built-in" information hinting at what a name means only has a historical interest once the name has become established. This is also referred to as the Juliet Principle (once again following Collins, 1977) by virtue of Juliet Capulet's much-quoted line "What's in a name?" in William Shakespeare's play *Romeo and Juliet* (2010 [1597]). What Juliet essentially says is that the name does not matter, the important thing is what it denotes and what people think and say about it.

  There is definitely an important truth to this. To figure out the full meaning of an unfamiliar word, children and adults alike need to rely on a wide range of surrounding cues spanning from oral explanations and written descriptions or definitions to first-hand visual and other sensory experiences and physical action (e.g., Bloom, 2002; Graves, August, & Mancilla-Martinez, 2013). Places are no exception: we hear about a place, we see pictures from it, and we may ultimately decide to go there to see (smell, feel, etc.) it for ourselves – thereby gaining a still better idea of what that place is.

However (as also noted by Collins, 1977, while neglected by some scholars), this does not mean that the Joyce Principle cannot still have a decisive bearing on the final outcome. What a name says literally can both guide and misguide a recipient when interpreting other (name-extrinsic) cues and this may result in both adequate and less adequate perceptions of the referent, in our case: a place (for experimental evidence from the related domain of place-related food names, cf. Smith et al., 2022; Smith, Barratt, & Zlatev, 2014). Moreover, such misconception may sometimes become permanent. For example, until told otherwise a research colleague of ours solemnly believed that the Danish regional specialty called *kålpølse* ('cabbage sausage') actually contained cabbage where the truth is that it is traditionally served with stewed cabbage in Sothern Jutland.

Instead of seeing the Joyce Principle and the Juliet Principle as oppositions, it is more informative to see them as complementary mechanisms that jointly influence the formation of (in our case) the understanding of a place name and a corresponding brand.

- *Perspective 4: The full ecosystem of naming and framing processes.*
  When it comes to wider topic, such as climate change, immigration, or sustainable tourism development, the goal-driven use of naming and framing rarely comes down to the use of a single name. Whole "sets" of names will typically come into play when interested stakeholders address a given topic, often presenting it in a very different light. When discussing tourism development, for example, some will speak of new job opportunities, improved infrastructure, and preservation of local culture and heritage. Others, however, will speak of overcrowding, seasonal unemployment, and environmental and cultural damage.

  When in play together, such words will mutually influence the understanding of each other and may well push it in opposite directions (take *climate change* which can be framed both as an urgent problem and as an entirely natural phenomenon). Some names are tailored from the outset to support a particular position which is sometimes encoded directly into the structure of the name, i.e. the Joyce Principle based part of its semantics, e.g., *ecotourism* versus **overtourism**. Others are subject to constant semantic negotiation driven by name-extrinsic cues, i.e. what we here call the Juliet Principle, say, about what exactly qualifies something as *sustainable tourism* when applying existing normative definitions to concrete instances; see also Collier, Hidalgo, and Maciuceanu (2006) for comparable examples viewed as instances of so-called essentially contested concepts.

  In sum, what we are dealing with is a whole ecosystem of naming and framing processes where the mechanisms described under Perspective 1–3

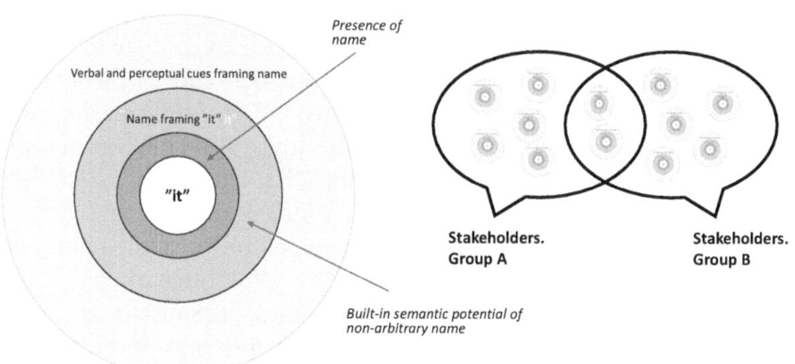


*Figure 3.1* The four dimensions of naming and framing visualized (figure created by the first author, slightly modified after Smith, 2021: 6).

also operate across the individual nodes of the system (see Smith, 2021: 62–65, for further discussion). Figure 3.1 illustrates the four perspectives and their interconnections in visual form.

However, even if Perspective 4 is clearly relevant to the INCULTUM agenda as a whole, in the present chapter, we will focus mainly on a single name, namely Campina de Faro, and its perception by visitors to the Algarve coast. For that reason, the analyses to follow will centre around the mutual dynamics between Perspectives 1and 3.

### 3.2.2   *Low- versus high-budget route brand development*

Some branding consultants, especially at the high end of the price range, tend to foreground (what we here call) the Juliet Principle at the expense of the Joyce Principle. The classic sales talk goes: "You can call your brand whatever you like! It can still be a great success if you have a great product, a solid marketing budget, and a top-tuned branding team behind you".

Needless to say, the situation is directly opposite for most INCULTUM pilots. The same is true for many other commercial and non-commercial endeavours, which means that the reverse challenge, branding on a budget, is also a central focus for some branding advisors (e.g., Mankoo, 2023; Tailor Brands, n.d.). However, the theoretical motivation for the advice given is often rather heterogeneous, if present at all, and only rarely encompasses factors of special relevance to place branding (such as an active role of the locals, cf. Freire, 2009; San Eugenio et al., 2019).

Taking the theorizing a step further, we here introduce a distinction between a *high-budget route* and a *low-budget route* of brand and product-name development originally suggested by Riezebos, Kist, and Koostra

(2003: 80–103) and taken further in certain respects by Smith (2021: 43–53). It incorporates Collin's (1977) distinction between the Joyce Principle and the Juliet Principle with a particular focus on how the corresponding variables engage in different mutual dynamics depending on the size of the branding budget.

### 3.2.2.1   *The high-budget route*

The high-budget route presupposes that the framing of a brand and/or product (e.g. place) name is backed up by extensive and costly communication efforts drawing on an elaborate mix of paid advertising, free media coverage (yet generated by spending substantial resources on promotion and PR, events etc.), active social-media presence intended to accelerate and shape seemingly self-driven communication processes, and the use of self-owned media platforms offering brand-supporting content. For an overview of possible mixes of such paid, earned, shared, and owned media strategies and the still more blurring lines between them, see Macnamara, Lwin, Adi, and Zerfass (2016).

Clear examples of the high-budget route are a corporate brand like *Apple*, including branded product lines like *iPhone* or *Mac*, or a major place brand like *Copenhagen*, along with branded attractions such as *Royal Arena*. Here, the Juliet Principle plays a predominant role from the outset in the shape of multiple verbal, visual, and experiential cues surrounding the name at multiple touchpoints across public space and cyberspace. In so far as the choice of name has been part of the branding effort (which is, of course, not the case with a historically given name like Copenhagen), the name's intrinsic semantic potential, i.e., the Joyce Principle, may add some freshness ("apple") or prestige ("royal"), but such contributions remain optional.

### 3.2.2.2   *The low-budget route*

In the low-budget route, by contrast, the brand image is crafted without significant advertising or other costly communication efforts. In this case, the immediate communicative potential of the brand or product name will often play a more active role in combination with such cues that can be directly extracted from the branded item itself and/or from its immediate surroundings.

An example is the use of product packaging design as the primary branding tool, as illustrated in Figure 3.2 (see also Riezebos et al., 2003: 85–89; Rundh, 2009; Klimchuk, & Krasovec, 2013).

The product on the left, a seaweed-based caviar substitute, has become a worldwide sales success with a minimum of resources spent on advertising or other cost-intensive marketing efforts. The built-in morphological motivations of the brand name *Cavi-art* and the product name *Seaweed Caviar* have thus been sufficient to intrigue recipients in a relevant way and the rest of

*Figure 3.2* The low-budget route as implemented in packaging design. Images used with kind permission from Cavi-Art and Smartbox Group.

the story is told by the familiar shape of the jar and the surrounding visuals (along with the product facts declared on the back, for those consumers who take the examination that far).

Here, the outcome thus relies on a subtle interplay between the Joyce Principle and the Juliet Principle unfolding in a "micro-cosmos" displayed before the very eyes of the viewer, with no need for any complementary cues offered at other touchpoints. A comparable dynamics can be found on the packaging to the right with the additional twist of illustrating the still more widespread trend to present services as "packages" (LaPlante-Dube, 2017; Flash-Hub, 2022).

As for marginal cultural tourism destinations, simply "boxing them" is, of course, not a viable path, at least not as an isolated measure (though "boxed" holiday packages are indeed among the services offered to tourists). However, an example of an approach conceived in a comparable vein is a series of campaigns promoting tourism to the Faroe Islands. Opting for a minimal budget, these campaigns rely solely on close-to-product (here: place) cues made accessible to the global public in an online format. Volunteer Faroese residents invite remote viewers into their daily lives, tell stories, translate special Faroese words, and even let themselves be remote controlled to show around their premises. Sheep, in turn, walk the landscape in their own pace with a camera mounted on their back, thus offering Google Sheep View for lack of Street View (supported by Google Maps). For further details, cf. Mensch (n.d. a, b) and Visit Faroe Islands (n.d.). A taste of all this is given in Figure 3.3.

A marked difference compared to the scenario of Campina de Faro is that the physical distances between the majority of potential visitors and the place

*Figure 3.3* Innovative low-budget online campaign for visiting the Faroe Islands. Photo credit: Kirstin Vang. Used with kind permission from Visit Faroe Islands.

in question requires the sense of "being there" to be re-created in the shape of an interactive online universe. By contrast, with about 4.8 million tourists visiting the Algarve coast per year, relevant bottom-up cues pertaining to Campina de Faro could easily be presented within their immediate physical surroundings, supplemented by direct experiences on the spot for those eventually inspired to go there.

To establish a first baseline for any future initiatives to increase nearby sun-and-beach visitors' awareness of and interest in visiting Campina de Faro, while keeping the budget moderate, an explorational pilot study was conducted in the Faro-Albufeira coastal area.

## 3.3   Pilot study

### 3.3.1   *Methodology*

The study was conceived and prepared by the WP7 team at Copenhagen Business School and carried out in Algarve from September 29th to October 2nd, 2023, by the authors in collaboration with the Portuguese pilot.

#### 3.3.1.1   *Purpose*

The purpose was twofold: (1) to assess the extent to which visitors to the Faro-Albufeira area had previously encountered the name "Campina de Faro" given that this is the first and a vital precondition for any more elaborate place-brand image to evolve; (2) to gain some indications of which spontaneous expectations and associations the name was capable of triggering in respondents who either had or had not encountered the name before, and of the possible sources of these. Connecting back to the theoretical framework on naming and framing presented earlier, we thus combined an interest in Perspective 1 (whether potential visitors had/knew a name for the place at all) with a joint interest in Perspectives 2 and 3 (the influence of the built-in potential of the name and of surrounding cues, respectively, on the understanding of it). The ultimate goal was to generate new leads about which name-intrinsic and name-extrinsic parameters could be operated on to strengthen tourists' awareness of and interest in visiting Campina de Faro while keeping costs moderate.

In the case of no pre-familiarity, the only possible source of influence would be the built-in cues (motivation) extractable from the name itself, i.e., the Joyce Principle. We here wanted to get an idea of the range of possible readings that could be brought about by the name as such and hence its potential as a first "icebreaker" for discovering and learning more about the place.

In the case of (some degree of) familiarity, the understanding of the name could have been influenced both by the above-mentioned, name-intrinsic factors and by extrinsic cues surrounding the name in such settings where it was first encountered and (if the case) re-encountered repeatedly later. In other words, it would be the result of an interplay between the Joyce Principle and the Juliet Principle, and we wanted to learn more about how that affected the understanding of the name as compared to the understanding displayed by participants who had never heard it before.

### 3.3.1.2   *Target population*

The intended target population was adult foreign and domestic visitors[1] to the Faro-Algarve coast. We thus opted out local residents as respondents, even if it might have been interesting to test out earlier indications of a low awareness of Campina de Faro even among the locals. As it were, we did get some indications of this through our less formalized field observations described below. For the survey, however, we maintained a clear focus on the scenario of prompting tourists staying by the nearby coast to explore the hinterland.

A homogeneous convenience sampling strategy was implemented (Jager, Putnick, & Bornstein, 2017) with an emphasis on ensuring that our respondents could mainly be regarded as tourists and as non-residents. Apart from that, a diverse sample of visitors of various nationalities was opted for to enrich and diversify the study's findings. The recruitment of participants primarily took place in the out- and indoor areas of hotels and at public beaches, as well as on some neighbouring locations, in the Faro-Albufeira area. This included Hotel Faro, Eva Seuses Hotel in Faro, Hotel Alisios in Albufeira, and beach sections in Faro and close to Albufeira. A few responses were furthermore collected at the market in Olhão, at the gate of Faro Airport, and on the aeroplane with tourists returning from Faro.

### 3.3.1.3 *Questionnaire survey*

To support the objectives outlined above, we designed a short questionnaire (entitled a "micro-survey"), including both quantitative and qualitative questions, which was printed on paper and filled in by the respondents by hand. To accommodate the multilingual target group, the questionnaire was prepared in an English version as well as a one in Spanish, Portuguese, French, German, Swedish, and Danish to prevent potential linguistic difficulties with understanding and answering the survey questions.[2]

The respondents were informed about the overall purpose and institutional context of the survey and gave their consent in advance. Responses were confidential and anonymous, and the respondents had the autonomy to conclude their involvement in the survey at any point. The survey design was authorized by the Data Protection Office of the University of Algarve.

The opening question was whether the respondents had ever heard the name "Campina de Faro". If they answered "yes", they would be directed to the next page and asked to use up to three words or short phrases to describe Campina de Faro according to their own knowledge or experience. They were likewise asked where they had first heard the name, as far as they could remember (open response), and whether they had ever been there (yes/no). The respondents who answered "no" to the first question were instead asked to use up to three words or short phrases to describe what they would *image* that "Campina de Faro" could be and skip the next page.

These questions were followed by demographic questions about home country, gender identity, and age, as well as reason for visiting the Algarve, place of stay, and total number of visits to the region. Finally, all participants were presented with a short description of Campina de Faro condensed from the pilot description on the INCULTUM homepage, including the photo depicting a waterfall featured there. They were then asked about the likelihood that they would consider visiting Campina de Faro or going hiking there. Participants were given the time necessary to fill out the paper sheets, and space to reflect, also aloud, on their answers which in some cases gave us fruitful additional insights.

### 3.3.1.4 *Field observations*

In addition to the questionnaire-based survey, less formal but valuable observations were made by us throughout our presence at the relevant locations (reached on foot or by car). This includes informal conversations with local residents, particularly such who were kindly rejected for the survey but still contributed with valuable comments and perspectives. We also gathered some input from employees in the tourism industry, including vendors of bus excursions, hotel staff, and the employees of a car rental company.

At the third day of our stay, we made a field trip accompanied by three fellow researchers from the Portuguese pilot team to better identify and learn about Campina de Faro ourselves as we had never visited that area before (or rather: we were not aware that we partially already had, see below). During the 3-hour tour, we endeavoured to see the most important spots of Campina de Faro – among them the orchards, old watermills, and aqueducts. The trip also gave us an idea about the size and fuzzy boundaries of the whole area. The observations made and notes taken contributed substantially to a wider contextualization of the results gained from the questionnaire survey and will be drawn on in the general discussion section. This also includes essential factual and terminological details provided by our colleagues.

### 3.3.2   *Survey results*

In total, we collected 92 valid filled-out questionnaires over the course of our four days on-site. Many more respondents were approached and willing to respond but turned out to be local residents enjoying beach life etc. alongside the tourists and were therefore kindly rejected.

The data were transposed to an excel sheet and subject to quantitative and qualitative analyses. Given the exploratory character of the study, the presentation of the findings below relies on descriptive statistics only. We had no intention to filter out any potentially statistically significant correlations across the whole data set or testing any pre-fixed hypotheses.[3] The purpose was to provide a versatile snapshot of tourists' perceptions (or lack of such) of Campina de Faro, thus enhancing the knowledge base for ongoing innovation actions and possible follow-up inquiries.

### 3.3.2.1   *Participant characteristics*

Out of the 92 respondents, 58 (63%) were female, 33 (35.9%) were male, and one did not want to state the gender, see Figure 3.4. Given the widely recognized tendency towards an overrepresentation of females in questionnaire survey responses (cf. Groves et al., 2009: 183–211), we considered this as an acceptable gender balance.

In terms of age, the participants were quite nicely distributed as well, with at least 14 participants in every age group defined by us, as shown in Figure 3.5.

As for home countries,[4] the sample was quite varied, see Figure 3.6. The majority of visitors recruited were from European countries, whereas 12 were from countries outside Europe (Ecuador, USA, Australia, New Zealand, and Canada). Among the Europeans, more than half were from either Germany, the UK, the Netherlands, or Portugal, while the rest were distributed across 11 other European countries in both North, South, East, and West Europe. The distribution aligns rather well with the official statistics of visitors to the Algarve Coast where domestic travellers are accounted for the highest number in 2022, with tourists from the UK coming in second (for details, see Luz, 2023).

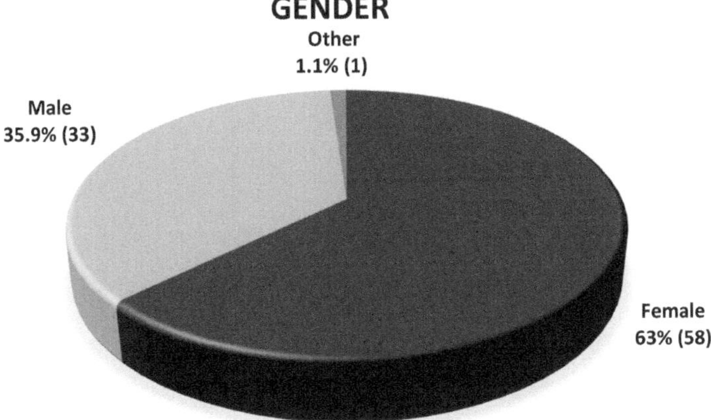


*Figure 3.4* Distribution of respondents by gender.

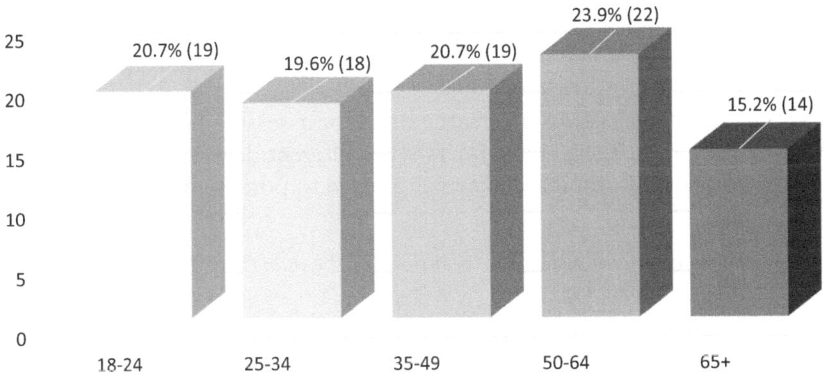


*Figure 3.5* Distribution of respondents by age.

While 50.5% of the respondents were first-time visitors to Algarve, 37.4% indicated that they had been in the Algarve 3 times or more. Only 12.1% were in the Algarve on their second visit. Out of the 92 respondents, 70 (76.1%) indicated vacation/holiday as the main reason for their visit. Other reasons included work/business, study/training, and visiting friends or relatives. However, given that the latter respondents were encountered in a tourism context (on the beach, waiting in line for a boat trip, etc.), we considered their responses as valid input for the current purpose.

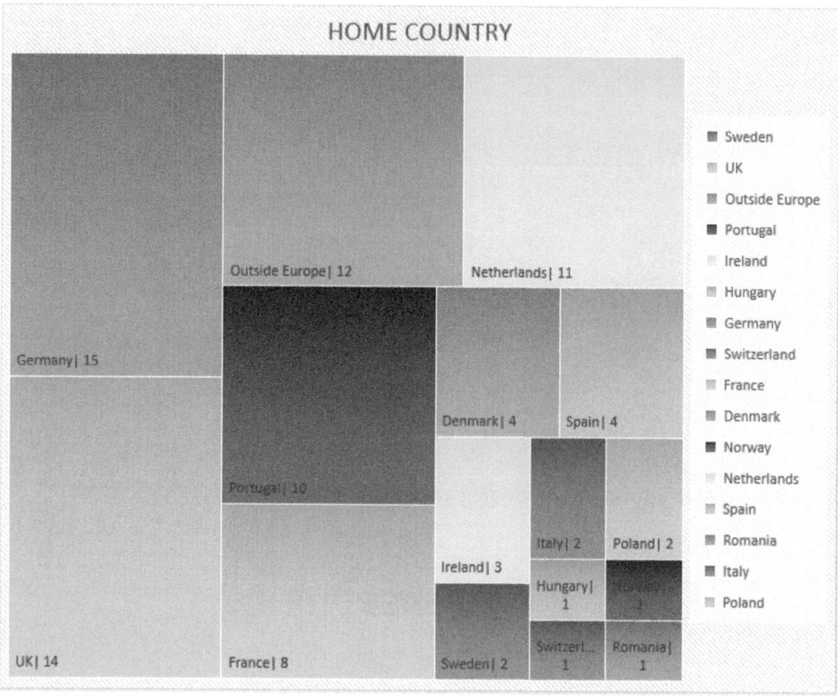


*Figure 3.6* Distribution of respondents by home country.

The sample thus gave us a versatile basis for assessing how visitors of different nationalities, ages, and genders with different degrees of experience of the area perceived Campina de Faro as a place to potentially visit.

### 3.3.2.2   *Familiarity with and comprehensions of the name "Campina de Faro"*

Our pre-expectation was that a substantial number of the respondents might not have heard the name "Campina de Faro" before, considering also of the tentative indications we had from the earlier inquiries mentioned initially. That expectation was more than confirmed, as shown in Figure 3.7.

Out of the 92 respondents, only seven (7.6%) declared to have heard the name before, which could either indicate that they knew the place or that they had merely stumbled upon the name at some point. For the remaining 85 respondents (92.4%), however, the first and crucial precondition for any more far-reaching place-branding efforts – namely people knowing a "name for it" as further discussed under Perspective 1 in the theoretical framework introduced earlier – was not present. Notably, the issue here is not that a well-consolidated name does not exists (as could be the issue, say, for a newly established company) but that the respondents did not know it.

## HEARD OF "CAMPINA DE FARO"

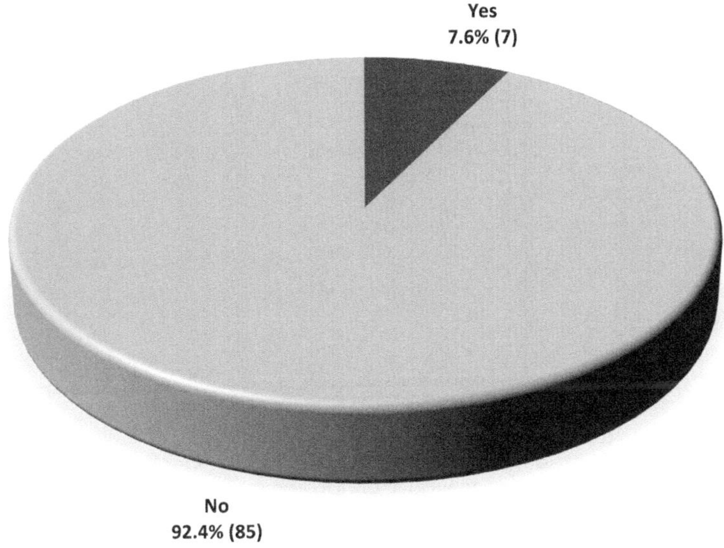


*Figure 3.7* Responses on ever having heard the name "Campina de Faro".

In turn, this means that any expectations and associations that the name might evoke in these respondents would be due to the name's own built-in communicative potential (motivation), i.e., the Joyce Principle as covered by Perspective 2 in the present theoretical framework. To ensure this, we abstained from offering any additional cues that could help the interpretations along at this stage. The keywords and phrases noted down by these respondents in an attempt to describe what they would *imagine* Campina de Faro to be were therefore a valuable indication of how the intrinsic semantic cues built into the name might actually be interpreted when first encountered in a real-life communicative setting, including such settings that could support low-budget brand building.

The free-text responses given by the 85 respondents in question were subject to a qualitative content analysis performed by the two authors. The aim was to filter out characteristic lines of interpretation and forming corresponding meta-categories which were subsequently discussed and cross-validated by the authors and finally consolidated. Figure 3.8 shows the top five of these meta-categories, i.e., those that materialized in the largest number of individual responses, with some variation in wording.[5]

The top five suggest that the intrinsic morphological motivation of the name was relatively transparent even to non-Portuguese speakers, implementing

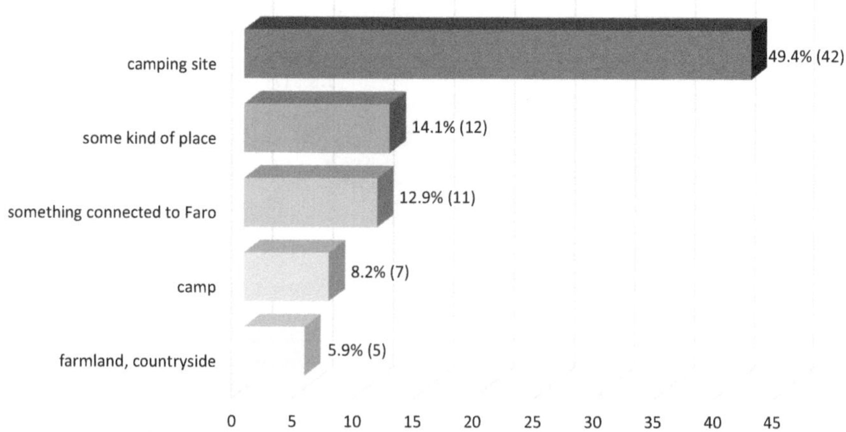



*Figure 3.8* Top five of spontaneous interpretation of "Campina de Faro" by participants who had not heard the name before, expressed as percentage of responses that contained elements matching the respective meta-categories.

the classic Romance-language N-de-N (noun + preposition + noun) construction with one word "Campina" (the head), being further specified in one way or the other by another word "Faro" (the modifier). Most participants dwelled at the head, drawing on cross-nationally well-known and comparable sub-senses of the Latin-origin word. The most frequent reading was "camping (site)" found in 49.4% of the responses, possibly inspired by the overall tourism context (though not quite adequate, at least for the area as a whole). Others thought of a "camp" (e.g., for kids) or of "farmland" or "countryside" which is indeed in line with what "campina" can also mean (along with "campagna" in Italian, "campo" in Spanish, etc.). Still others dwelled at the modifier "Faro" and suggested that it had to be "something" connected to Faro (with more specific suggestions found at the below-top-five level, e.g., the "castle of Faro" or "a campaign for Faro"). Notably, across all suggestions there was a wide consensus that it must be some kind of place, as captured in pure form by reading number two in the top five (found in 14.1% of the responses) which covers suggestions simply indicating "some" place as expressed by words like "neighbourhood", "district", "area", or "landscape". Below the top five, again, we find more precise suggestions such as an "airport", a "golf court", or a "park".

In sum, the name "Campina de Faro" indeed does seem to possess the intrinsic qualities needed to make it a suitable pivot for building a more precise and elaborate place-brand image. However, some additional help from surrounding, name-extrinsic, cues is clearly required to push people's interpretations of the name further in a desirable direction. Stated in the terms introduced earlier: the Joyce Principle (covered by Perspective 2) does a fine job, but the Juliet Principle (covered by Perspective 3) needs to get involved as well for people to generate any more elaborate perception of what Campina de Faro really is.

We had aimed to learn more about which kinds of name-extrinsic cues that had contributed such an outcome for those respondents who declared to know the name already by asking them to indicate *their* perception of "Campina de Faro" (in keywords) as well as where they had first heard the name, as they remembered, and whether they had visited the location themselves. The input on that point was, however, quite scarce, with only seven respondents to compare. Four of these (three Portuguese and one UK citizen) seemed to have a quite clear perception of the place, mentioning different but all relevant keywords such as "rural", "dry", "flat", "farmland", "peasants", "countryside", and "food and drink". One had heard the name as a child, another at work, while the third didn't remember and the fourth skipped the question. Only two of them had actually been there. This gives at least some indication that the name can come to be understood in a desirable way through surrounding communicative framings alone, even without yet having had any on-site experiences. By contrast, the three other respondents (three Dutch first-time visitors) who answered "yes" to having heard the name gave less applicable keywords (including "camping") and no details about how/where they heard it. They might well simply have misinterpreted the question.

### 3.3.2.3 *Intentions to visit after the "reveal"*

A very clear indication of what the Joyce Principle and the Juliet Principle are capable of accomplishing when combined did however follow from the responses to the concluding part of the questionnaire, the "reveal" part. Here, the actual reference of the name "Campina de Faro" was made more clear to the respondents via additional verbal and visual cues in the shape of a condensed version of the description featured on the INCULTUM homepage, including the photo.

When asked if they would consider visiting Campina de Faro based on this short description, 78 answered "yes", 11 answered "not sure", and only one answered "probably not" (while two skipped the question), see Figure 3.9.[6] Moreover, a substantial number of participants got quite excited about the idea of going to the place and asked us many questions about how to get there, what were the most characteristic things to see, and so on.

This is a clear indication that very simple Juliet-Principle–type cues such as a short text and a photo can make a tremendous difference when it comes to shaping people's understanding of the name and creating a desire to go to the place and find out more. That is, it lays the ground for the emergence of a positive place-brand image that can be subject to continued development. Moreover, descriptions comparable to the one we used do not need to be widely circulated through costly advertising or other mass communication.

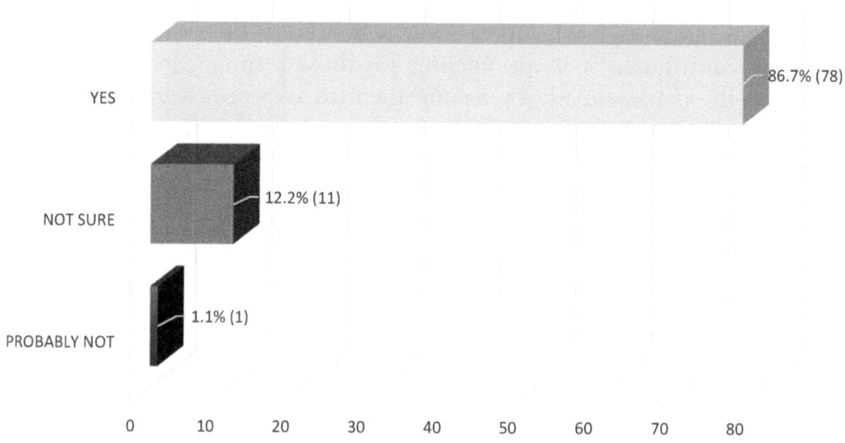


*Figure 3.9* Respondents' intentions to visit Campina de Faro after having read a short description of the area.

They could be made available, say, as leaflets, on posters, etc., in places where tourists spend time, including relevant websites apart from INCUL-TUM's own.

However, the respondents' eager interest and questions – combined with the additional information gained through our own field observations in the area and especially the trip around Campina de Faro – also gave us some new food for thought. During the first two days, we had to openly admit that we could not guide them much further since we had not as yet been there ourselves. When continuing the data collection after the field trip, we additionally felt that answering the questions was less straightforward than the participants (and we) might have expected. We now knew that figuring out where exactly to go and what exactly to see in the extensive area named Campina de Faro could be a challenge in itself, and that this could hardly be done in a single trip. The wider implications of this insight lead us to the general discussion and considerations about possible next steps.

## 3.4   General discussion and next steps

### 3.4.1   *From image to identity, and back*

Our survey contributed essential insights on the place-brand image aspects, i.e., on how potential visitors see the place, whereas our field research gave us some additional indications of the underlying place-brand identity, i.e., on how the locals see the place themselves, which also includes the physical manifestations of this in the built environment, monuments, etc. (cf. Kaefer, 2021: 7).

Combining both perspectives, we gradually became aware of two additional, and closely connected, challenges when it comes to strengthening the awareness of Campina de Faro as a potential target for cultural tourism.

The first is the size, fuzzy boundaries, and heterogeneous character of the place. It stretches from the Atlantic Ocean in the South between the cities of Faro and Olhão in the East and Loule and further to Albufeira in the West and all the way up to the mountains in the North which makes it an area a lot larger than we expected from the narratives heard earlier. The landscapes include the *litoral*, the flat area close to the Algarve Coast, the *campina*, the hilly and dry farmland between the ocean and the mountains (and also eponym for the whole place), the *bacoral*, defined as the foothills of the mountains (*serra*) to the North used mainly for herding. The potential tourism attractions range from the ancient water structures from the Islamic period which are scattered across the whole area (in varying state of maintenance and some located in private front gardens and courtyards), a mix of traditional and modern farmland, and ancient architecture (often encapsulated by more modern), to a captivating waterfall. Similarly, individual towns and villages within the area, such as *Estoi, Patacão,* and *Conceição*, each possesses their own distinct characteristics, identities, and rich histories. Significant portions of Campina de Faro have furthermore evolved beyond a peripheral designation, now hosting motorways, residential areas, shopping malls, and a large football stadium.

In sum, there seems to be grounds for arguing that Campina de Faro does not quite constitute a destination for tourist visits in its own right, but rather several and very different potential sub-destinations that could hardly be covered by a single trip. That was also mirrored in our own on-site experiences: During our first days of data collection, we drove through and past parts of Campina de Faro plenty of times without being aware of it – while looking forward to finally "get there".

The second challenge is that not only tourists but also locals seem to have a limited awareness of Campina de Faro understood as "one place", and indeed of the very name. During our many casual chats with local people, including potential respondents for the survey who had to be rejected because they were local, we brought up the name Campina de Faro and asked whether they could tell us more about it. In many cases, people simply admitted that they had no clue, while some locals did in fact know the name but explained that it was basically not a particular place but just "all of this", pointing from the beachside towards the hinterland. That includes Portuguese working within the tourism industry such a car rental lady and a seller of bus tours (and colleagues they consulted) who at best also landed on something like "ahh, it's all of this". In addition to this, there were no road signs, no labels, and no other haptic touchpoints that could be attributed to Campina de Faro, at least as far as we could observe when moving around in the area, and as later confirmed by the Portuguese partners.

In sum, the availability of close-to-product (here: place) cues that could introduce the name and further shape the understanding of it by expanding on its intrinsic semantic potential, whether in the shape of word of mouth from locals (which has proved to be an essential factor in other scenarios, cf. Freire, 2009; San Eugenio et al., 2019) or on-site written or visual cues, was close to zero.

### 3.4.2    *What to do?*

While we are not in a position to give any definite advice or recommendation, we will still briefly summarize some considerations that might prove useful for ongoing and coming actions undertaken by the Portuguese pilot on the backdrop of our findings. The most salient consideration is probably whether Campina de Faro (and its name) should be retained as the main focus for the promotion of cultural tourism in the hinterland of Faro, or whether a more narrowly defined focus point, or a number of such, should be opted for in the long run.

In so far as the status of Camina de Faro as a "master brand" is retained, it would probably be vital to enhance that name's visibility at key touchpoints such as tourism offices, tour providers, hotels, restaurants, and relevant websites, in combination with easily comprehensible written and visual information about Campina de Faro and (not least) its many specific attractions. The strong effect of such simple measures bringing the Joyce Principle and the Juliet Principle into a subtle interplay with each other in a close-to-product (here: place) context was clearly demonstrated by our survey results: 86.7% of participants were ready to visit the area after taking the survey, even if only 7.6% had ever heard of it before. Apart from printed input like that included in the survey, vital factors could also be the spreading of comparable cues in oral form, not least by people working in the tourism industry, and remedying the lack of road signs, info posters, etc.

However, as we saw it earlier, the challenge here is not only to provide a name for a place but also to provide a suitable place to go to once the name has caught peoples' interest. In that light, narrowing down the focus to selected spots within the Campina de Faro area, perhaps just 2–3 for a start, could be a viable path. In that case, communicative measures similar to those just mentioned could be applied, yet starting directly from the (names of the) spots selected. In some cases, this might well include the need to create a more "tourism-friendly" (nick)name for the place or sight in terms of the name's built-in communicative potential and immediate comprehensibility to the target audiences.

Indeed, we were introduced to one suitable target spot for such an approach during our field trip: an old historical farmhouse and a water-mill now located in a major traffic hub close to the relatively newly-built

*Teatro das Figuras* which dominates the space visually. Yet that theatre could potentially also offer an excellent starting point for its visitors (and for others) to acquaint themselves with the nearby historical heritage, provided that signs would point them in the right direction and posters or even human guides on the spot could tell them more about their intriguing history. As part of that, instead of just speaking of "an" old farmhouse and "a" water mill, as we just did, more salient names could be (re)discovered and/or created for these attractions for the purpose of tourism promotion. Moreover, guided tours to other locations in Campina de Faro could be offered from such a spot, sold by tour operators there or even inside the theatre. A different option would be to focus on particular villages selected for their specific identity, history, architecture, etc., brought to life, for instance, by storytelling offered by older-age citizens volunteering as local guides.

### 3.4.3   Summing up

Returning to RQ1, the name "Campina de Faro" is evidently underexploited as a possible fixpoint for developing a more salient place-brand image among potential visitors, as showcased by the near-to-complete unawareness of the name (and place) among our survey respondents. However, the answer to RQ2 is that both name-intrinsic and name-extrinsic parameters do in fact exist that can be operated on to remedy the situation. The information built into the name itself was thus capable of eliciting quite adequate expectations in most respondents, excluding any need for re-naming the place for tourism purposes. In turn, a more elaborate, and predominately positive, understanding of the name could be achieved by surrounding it by simple close-to-product (here: place) cues, as demonstrated by the short text and photo included in our own questionnaire. What seems more challenging is to match the area itself with the positive expectations thus evoked. That may be easier done piece by piece, focusing on particular attractions within that area.

### 3.4.4   Beyond Campina Faro: cross-pilot learnings

A major priority of INCULTUM is to facilitate a dynamic exchange of ideas, practices, and learnings across the participating pilots and other marginal and remote tourism destinations. This takes us further to RQ3. What our inquiries might first and foremost contribute at a cross-pilot level are certain new insights into the specifics of place-branding in what we have earlier referred to as a "below zero" scenario. That includes the advantages and limitations of applying a low-budget brand development route relying on such cues that can be directly extracted by visitors from the targeted area itself and/or from its immediate surroundings.

While not all INCULTUM pilots display such a "below zero" scenario to the full, other examples that do seem to qualify include the Albanian pilot targeting the Upper Vjosa Valley bordering to Greece, and the Sicilian pilot targeting the inland territories of the Trapani Mountains. In both cases, the area in focus covers a vast geographical territory and the potential tourism attractions are rather diverse with some also located at a substantial distance from each other, posing challenges also of transport accessibility. Furthermore, like in several other INCULTUM pilots, current tourism flows are limited, the stakeholder alliances needed for developing and operationalizing seemingly promising cultural tourism opportunities are not yet fully established, and the financial resources available for the purpose are scarce.

Learnings from the present inquiry which could (with some adaptations) also be of relevance for these cases include a clearer recognition of the schism between promoting the area as a whole and finding suitable "front-runners" within that area that are easier to name, frame and ultimately "sell" to tourists. Another relevant point would be the emphasis on finding suitable touchpoints and affordable communicative tools for addressing potential visitors in their immediate surroundings (and on the spot, for those who eventually go there) rather than relying on more traditional marketing tools. That includes recognizing how effective such low-budget approaches can actually be, as clearly demonstrated also by our own survey: The vast majority of respondents were quite determined to go and visit Campina de Faro at some point after having read a short description and seen a photo, regardless that most of them had never even heard the name before.

Conversely, there might be some learnings for Campina de Faro to gain from the two other pilots just mentioned. In terms of concentrating the focus on selected spots (among many good candidates), the Albanian pilots opted for concentrating their resources on restoring the heritage of the ethical minority known as the Vlachs who earlier lived as nomads in the area. Specifically, the Albanian team came up with the idea to install an old Vlach dwelling that could work as a centre of attraction for tourists to learn about the Vlach way of living while also providing a good base for further exploring the surrounding territories via activities such as camping and hiking. In this way, more than one important aspect of the region can be unified into one tourist attraction. As for the Trapani Mountains, it is planned to convert an abandoned railway line crossing a beautiful landscape into a hiking trail, which, in turn, can also lead the hikers to other key sights in the area. In this case, the idea is furthermore backed up by innovative naming and framing in that the route has been given its own name: the Green Line.

The conceptual framework and empirical findings presented in this chapter, of course, remain minor contributions to meeting the multiple and diverse challenges facing the development of cultural tourism in Campina de Faro, Upper Vjosa Valley, and Trapani Mountains, as well as in the remaining INCULTUM pilots. Nevertheless, we hope that they might have opened a few new perspectives that can enhance future debates on how to best tackle some of these challenges, and corresponding actions.

## Notes

1 We consistently declined interested respondents who turned out to be residents of Algarve and not visitors. However, borderline cases were bound to exist, such as foreigners living (semi-)permanently in Algarve or locally born people currently on vacation, and a few such instances may well have slipped through. We do not see this as a major methodological concern given that the vast majority of respondents, regardless of national background and place of permanent residence, responded "no" to having ever heard the name Campina de Faro and answered the follow-up questions accordingly.
2 Qualitative responses in other languages than English were subsequently translated to English (in being our working language) by us and colleagues with proficiency in the relevant languages.
3 We did, of course, have certain pre-expectations, such as a limited awareness of Campina de Faro among tourists and a high degree of variation in their spontaneous understanding of the name, but these could be clearly confirmed by simply looking at the naked data.
4 In this and a few other cases, one or two respondents accidentally skipped the question. In such cases, any percentages are based on the number of responses who actually responded to the question.
5 Some respondents contributed to more than one of the top-five categories and others to none, given that many respondents listed two, three, or even more different suggestions among which some would be frequently recurring across participants (like "camping site") and others more idiosyncratic.
6 When additionally asked if they would also consider going hiking in Campina de Faro, 62 (68.9%) answered "yes", 14 (15.6%) answered "not sure" and 14 (15.6%) answered "probably not" (while two skipped the question once again). Some of the more reluctant participants spontaneously commented on their response, the typical "excuse" not being a lack of interest in visiting the place but in hiking as such, difficulty walking long distances due to old age, etc.

## References


Barsalou, L. W. (1987). The instability of graded structure: Implications for the nature of concepts. In U. Neisser (Ed.), *Concepts and conceptual development: Ecological and intellectual factors in categorization* (pp. 101–140). Cambridge University Press.

Barsalou, L. W. (2016). Situated conceptualization: Theory and applications. In Y. Coello & M. H. Fischer (Eds.), *Foundations of embodied cognition, Volume 1: Perceptual and emotional embodiment* (pp. 11–37). Psychology Press.

Batista, D. (2023). The role of participatory-collaborative approach in communnity-based cultural tourism. The case of Portuguese pilot of INCULTUM project. *Paper presented at the ERSA2023 62nd Congress.* University of Alicante.

Bloom, P. (2002). *How children learn the meanings of words.* MIT Press.

Boisen, M., Terlouw, K., Groote, P., & Couwenberg, O. (2018). Reframing place promotion, place marketing, and place branding – moving beyond conceptual confusion. *Cities, 80,* 4–11.

Bussmann, H., Kazzazi, K., & Trauth, G. (2006). *Routledge dictionary of language and linguistics.* Routledge.

Collier, D., Daniel Hidalgo, F., & Olivia Maciuceanu, A. (2006). Essentially contested concepts: Debates and applications. *Journal of Political Ideologies, 11*(3), 211–246.

Collins, L. (1977). A name to conjure with: A discussion of the naming of new brands. *European Journal of Marketing, 11*(5), 337–363.

Entman, R. M. (1993). Framing: Toward clarification of a fractured paradigm. *Journal of Communication, 43*(4), 51–58.

Flash-Hub. (2022). What is service productization and how to productize our service. *FLASH-HUB Services.* https://www.flashhub.io/service-productization/

Freire, J. R. (2009). 'Local People' a critical dimension for place brands. *Journal of Brand Management, 16*(7), 420–438.

Graves, M. F., August, D., & Mancilla-Martinez, J. (2013). *Teaching vocabulary to English language learners.* Teachers College Press.

Groves, R. M., Fowler Jr, F. J., Couper, M. P., Lepkowski, J. M., Singer, E., & Tourangeau, R. (2009). *Survey methodology* (Vol. 561). John Wiley & Sons.

INCULTUM. (n.d.). *Public délivrables.* URL: https://incultum.eu/deliverables/

ISO. (2021). ISO 20671-1:2021(en): *Brand evaluation—Part 1: Principles and fundamentals.* ISO. https://www.iso.org/standard/81739.html.

Jager, J., Putnick, D. L., & Bornstein, M. H. (2017). II. More than just convenient: The scientific merits of homogeneous convenience samples. *Monographs of the Society for Research in Child Development, 82*(2), 13–30.

Kaefer, F. (2021). *Insider's guide to place branding.* Springer International Publishing.

Klimchuk, M. R., & Krasovec, S. A. (2013). *Packaging design: Successful product branding from concept to shelf.* John Wiley & Sons.

Kotler, P., Keller, K. L., & Brady, M., Goodman, M., & Hansen, T. (2016). *Marketing management.* Pearson.

LaPlante-Dube, S. (2017). How to market services – treat them like a product. *Precision Marketing Group.* https://www.precisionmarketinggroup.com/blog/bid/76722/how-to-market-services-treat-them-like-a-product

Luz, B. (2023). Number of tourists to the Algarve 2022, by country of origin. *Statista.* https://www.statista.com/statistics/1155164/number-of-tourists-to-the-algarve-in-portugal-by-country/

Macnamara, J., Lwin, M., Adi, A., & Zerfass, A. (2016). 'PESO' media strategy shifts to 'SOEP': Opportunities and ethical dilemmas. *Public Relations Review, 42*(3), 377–385.

Mankoo, G. (2023). 12 easy steps for affordable branding on a budget. *LoGo.* https://logo.com/blog/branding-on-a-budget

Maurya, U. K., & Mishra, P. (2012). What is a brand? A perspective on brand meaning. *European Journal of Business and Management, 4*(3), 122–133.

Mensch (n.d. *a*). Remote tourism. *Mensch.* Place Branding. https://mensch.dk/uk/cases/visit-faroe-island-remote-tourism-2/

Mensch (n.d. b). Closed for maintenance. *Mensch.* Place Branding. https://mensch.dk/uk/cases/visit-faroe-island-closed-for-maintenance/

Murphy, G. L. (2010). What are categories and concepts? In D. Mareschall, P. C. Quin, & S. E. G. Lea (Eds.), *The making of human concepts* (pp. 11–28). Oxford University Press.

Nöth, W. (1995). *Handbook of semiotics.* Indiana University Press.

Oh, T. T., Keller, K. L., Neslin, S. A., Reibstein, D. J., & Lehmann, D. R. (2020). The past, present, and future of brand research. *Marketing Letters, 31*, 151–162.

Peighambari, K., Sattari, S., Foster, T., & Wallström, Å. (2016). Two tales of one city: Image versus identity. *Place Branding and Public Diplomacy, 12*(4), 314–328.

Riezebos, H. J., Kist, B., & Kootstra, G. (2003). *Brand management: A theoretical and practical approach*. Pearson Education.

Rugaard, N. (2022). *Attracting Danish tourists to Portugal: A case study of Campina de Faro*. Master Thesis. Copenhagen Business School.

Rundh, B. (2009). Packaging design: creating competitive advantage with product packaging. *British Food Journal, 111*(9), 988–1002.

San Eugenio, J. de, Ginesta, X., Compte-Pujol, M., & Frigola-Reig, J. (2019). Building a place brand on local assets: The case of the Pla De L'estany district and its rebranding. *Sustainability, 11*(11), 3218.

Shakespeare, W. (2010 [1597]). *Romeo and Juliet*. Edited by J. O'Connor. Longman.

Smith, S. (2015). A sense of place: Place, culture and tourism. *Tourism Recreation Research, 40*(2), 220–233.

Smith, V. (2021). *Naming and framing: Understanding the power of words across disciplines, domains, and modalities*. Routledge.

Smith, V., Barratt, D., & Zlatev, J. (2014). Unpacking noun-noun compounds: Interpreting novel and conventional foodnames in isolation and on food labels. *Cognitive Linguistics, 25*(1), 99–147.

Smith, V., Barratt, D., Møgelvang-Hansen, P., & Wedel Andersen, A. U. (Eds.), (2022). Study 4: "Local" by facts or by atmosphere? In *Misleading marketing communication: Assessing the impact of potentially deceptive food labelling on consumer behaviour* (pp. 79–98). Cham: Springer International Publishing.

Stock, F. (2009). Identity, image and brand: A conceptual framework. *Place Branding and Public Diplomacy, 5*, 118–125.

Tailor Brands. (n.d.). 7 tips for branding on a budget. *Taylor Brands Blog*. https://www.tailorbrands.com/blog/branding-on-a-budget

UAlg. (2023). *Perception of cultural trails/water heritage routes*. University of Algarve.

Ullmann, S. (1962). *Semantics: An introduction to the science of meaning*. Blackwell.

Visit Faroe Islands. (n.d.). *Remote tourism*. Press Material. https://visitfaroeislands.com/en/about1/marketing-development-campaigns/remote-tourism

Vuignier, R. (2017). Place branding & place marketing 1976–2016: A multidisciplinary literature review. *International Review on Public and Nonprofit Marketing, 14*(4), 447–473.

# 4 Innovative business models for cultural tourism

## Advancing development in peripheral locations

*Carsten Jacob Humlebæk and
Esben Rahbek Gjerdrum Pedersen*

### 4.1 Business models and business model innovation

Multiple definitions of business models exist but can in general be defined as: "the firm's value proposition and market segments, the structure of the value chain required for realizing the value proposition, the mechanisms of value capture that the firm deploys, and how these elements are linked together in an architecture" (Saebi et al., 2017, p. 567). Business models help scholars and practitioners to understand how an organization creates, delivers, and captures value from their products and services. Over the years, scholars and practitioners have identified, and labelled, a wide variety of business models with distinct combinations of value creation, value delivery, and value capture. For instance, Oliver Gassman and colleagues presented 55 business models in their book *Business Model Navigator*, including peer-to-peer platforms (e.g. eBay), open business models (e.g. Linux), the long tail (e.g. Amazon), and Razor & Blade (e.g. Gilette) (Gassmann et al., 2014).

The business model terminology emerged in the 1990s and was originally used to explain the rise of dotcom businesses which quickly replaced traditional, bricks-and-mortar companies with online platforms. However, in the 2000s, business model thinking gradually began to spread into the strategy, management, and organization literature (Chesbrough, 2007, 2010; Teece, 2010). In 2010, Alexander Ostervalder and Yves Pigneur's "Business Model Generation" further contributed to the popularity of business model thinking and made their "business model canvas" a commonly used tool in business schools, companies, and management consulting. The book was inspired by Alexander Osterwalder's (2004) doctoral thesis, which analysed digital business models. Today, the business model thinking is fully integrated into mainstream management theory and practice. Moreover, where business model theory and practice originally focused on commercial enterprises, the perspective is now used to analyse a wide range of organizations, including public agencies, non-profits, and hybrid organizations. The business model perspective has also been applied in the literature

DOI: 10.4324/9781003422952-4

on tourism and cultural heritage (Gatelier et al., 2022; Russo-Spena et al., 2022; Reinhold et al., 2017).

Business model innovation is about developing and implementing new ways for creating, delivering, and capturing value (Bocken & Geradts, 2020). Business model innovation can be the development of a brand new business model by a new or established organization or the change of a business model by an existing organization. For instance, Geissdoerfer et al. (2016, p. 1220) argue that business model innovation: "describes either a process of transformation from one business model to another within incumbent companies or after mergers and acquisitions, or the creation of entirely new business models in start-ups". Some discussions exist on how innovative business model innovation should be (how new should a new business model be?). However, it is fair to say that business model innovation moves beyond traditional improvements, e.g. optimization of existing technology, additions to the existing product portfolio, and improvements of existing service offerings. Business model innovation are more deep-rooted changes that require organizations to explore alternative architectures for creating, delivering, and capturing value to stakeholders.

## 4.2    The emergence of new, alternative business models

The concept of value is central to the understanding of business models, even though the concept is often ill-defined in the literature (Neesham et al., 2023). In the beginning, value was understood from a conventional business perspective, emphasizing the commercial benefits for customers and the company. Gradually, however, the literature increasingly focused on alternative forms of business models, which integrated both commercial and societal concerns. A plethora of concepts has been used to label these business models, including sustainable business models (Ringvold et al., 2023), circular business models, (Lüdeke-Freund et al., 2019), triple-layered business models (Joyce & Paquin, 2016), social business models (Yunus et al., 2010), sufficiency-based business models (Beulque et al., 2023), and collaborative business models (Pedersen et al., 2020). What these business models often have in common is (1) a more holistic view of value, (2) a broader perspective of stakeholders, and (3) a long-term horizon (Bocken & Geradts, 2020; Mignon & Bankel, 2023).

The design and implementation of alternative business models is not without challenges (Bocken & Geradts, 2020; Vermunt et al., 2019). Internally, the existing organizational structure and culture can inspire inertia which makes it difficult to discover and implement new, alternative business models (Kirchherr et al., 2018; Pedersen et al., 2018, 2019). Moreover, once established, there is a risk of "mission drift" where the organization increasingly become focused on commercial objectives at the expense of social and/or environmental goals (Ebrahim et al., 2014). Externally, the development of alternative business models often depends on active collaboration with external business partners, which may not have the necessary commitment or competences (Pedersen et al., 2019; Vermunt et al., 2019). Moreover, a new, alternative business model may, at least in the short term, come with higher costs that the customers are unwilling to pay for.

## 4.3   Contracts and business model innovation

Innovation does not only involve the products and services flowing between the stakeholders in a business model. The product and service offerings are dependent on the underlying infrastructure that allows these transactions to take place. Innovating the infrastructure can therefore also mean innovating the business model. To give a few examples, financial exchanges between the stakeholders are necessary to keep the business model afloat. Moreover, various types of information (product descriptions, shipping documents, contracts, invoices, guarantees etc.) are exchanged between the stakeholders, which are central for ascribing rights and responsibilities. Last, both finance and information flows depend on an underlying digital infrastructure which ensures connectivity between the actors. Innovating business models can therefore also mean looking at the underlying infrastructure enabling value creation, value delivery, and value capture. Moreover, developing new ways of how finance and information are managed and organized can in itself be a source of business model innovation.

In this chapter, we will focus specifically on the formal and informal contracts, which tie stakeholders together in business models. In recent years, we have seen a wave of new contract forms and finance mechanisms, which offer alternatives to the dominant business models in the market. For instance, the emergence of pay-for-success, value-based, and outcome-based solutions, which all reward impacts rather than activities, services, and products (Carter & Ball, 2021; Savell & Airoldi, 2020). What all of these models have in common is a dissatisfaction with current performance criteria, which is said to work against desired public and private goals. For instance, the healthcare system has been criticized for focusing too much on treating diseases rather than preventing them.

A good example of new contract forms and finance mechanism is the emerging field of social impact bonds (SIBs), which have gained momentum in the last decade (La Torre et al., 2019; Andersen, 2023). A SIB can be defined as: "an innovative financing mechanism in which governments or commissioners enter into agreements with social service providers, such as social enterprises or non-profit organizations, and investors to pay for the delivery of pre-defined social outcomes" (OECD, 2016, p.4). The SIBs thus bring together actors in a new form of cross-sector collaborative business model and change the fundamental logic of public service provision. However, some of the new contract forms and financing mechanisms have shown to experience significant implementation challenges, e.g. high transaction costs, inadequate data infrastructure, and lack of organizational competencies (Andersen, 2023). From a contracting perspective, it remains more complex to reward multiple stakeholders for collaborating on long-term impacts than compensating a single stakeholder for activities provided or units sold.

An underlying problem of new forms of contracting is linked to the concept of value. Even though it is possible to design a business model based on a more holistic perspective of stakeholders and value, it can be difficult to make contracts that accurately account for the value created and the contributions of the stakeholders. For instance, it might be that it is difficult to estimate the cultural value of a historical city, and the costs and benefits from, e.g. cultural tourism, may differ in time and be unevenly distributed among the stakeholders who ensure its maintenance and development (e.g. individual citizens, community groups, tourism businesses, municipalities, government, etc.). At worst, failure to align the interests can create stakeholder tensions as reflected in the recent local protests against mass tourism at popular destinations. Therefore, a business model requires formal and informal contracts which ensure a fair distribution of benefits between the network of stakeholders involved in the value creation. Some of the challenges, dilemmas, and solutions will be described in the following case from Granada. Here, the province of Granada has experimented with new, innovative payment-for-service contracts, which is intended to secure water supply, preserve historical heritage, and promote cultural tourism along the traditional irrigation systems.

## 4.4    Payment-for-service contracts in villages of the province of Granada (Spain) with traditional irrigation systems

In a number of villages and towns of the province of Granada,[1] a series of payment-for-service-contracts have been formalized or are being negotiated at the time of writing as a result of the INCULTUM project's impulse to develop a sustainable cultural tourism offer with its point of departure in the cultural heritage represented by the traditional irrigation systems. This case-based part of the chapter is based on the experience from Granada, where the INCULTUM team from the University of Granda has served as facilitator in the processes of negotiating the payment-for-service contracts.[2]

The principal stakeholders of these contracts are, on the one hand, the irrigators' communities that have been responsible for these territorial commons since times immemorial and, on the other, the town councils, that are to benefit indirectly from the increased activity following the sustainable increase in cultural tourism. There already exists an informal relationship between both institutions and, in many cases, a prior collaboration, which however does not have a legal framework. In fact, this collaboration in many cases takes place in a sort of a-legality, because, for example, when the town council performs work on an irrigation ditch in support of the community, it does so on a property that is not public. The contracts thus formally recognize, in many cases for the first time, a service that the irrigators' communities have always provided namely a stable supply of drinkable and irrigation water to those living in the municipalities in question as well as those living in other municipalities downstream due to aquifer recharge. But these ancient water management infrastructures also have multiple environmental, cultural, and economic functions and the contracts are also aimed at recognizing the ecosystem services that the irrigators' communities and the irrigation systems provide to the villages and the common good. The arid zones where these communities are situated are particularly subject to the effects of climatic change, which, however, are mitigated by the irrigation systems. They regulate the hydrological cycles increasing the availability of water in the basin through aquifer recharge thus also increasing soil fertility. The irrigated areas maintain a highly diverse agricultural production and provide green, ecological corridors that help to maintain a high biodiversity and have significant climate regulation effects providing cooler local climate and higher humidity. In doing so, these areas also serve as fire breakers in the case of wildfires and ensure availability of water for fire extinction.

Apart from these ecosystem services, the irrigation infrastructures are commons with a long historical tradition dating back, at least, to medieval times and thus play an important role in the culture and history of many regions. They are deeply rooted cultural systems which apart from distributing water in an equitable way and maintaining the common good through participatory and democratic structures are also indispensable for understanding the shaping of the cultural landscape and local identity.[3]

There is nothing new in this, which is actually part of the point. These hydrological systems with their accompanying cultural and social infrastructure are precisely part of a living heritage, but their value has often been neglected in recent decades due to modern agricultural trends which in these areas are based on pumping up water from underground aquifers. But what has been argued by agricultural consultants to be a more efficient use of water than the traditional irrigation systems has precisely proven to be unsustainable since modern agriculture drains aquifers without recharging them and thus in the medium term reduces soil fertility ultimately favouring desertification in these arid areas. The battle over water resources is increasing in many places in Spain and in other countries in Southern Europe these years. The recent controversy regarding the strawberry farming based on pumping water from underground aquifers on the borders of the internationally important Doñana National Park, in Southern Spain with devastating effects on the protected wetlands in the national park is a case in point.[4] The aquifers are being drained and the water levels are dropping, but similar dynamics around sustainable vs. unsustainable water use is at play in many other places.

In this perspective of sustainable water management, the traditional irrigation communities have received a renewed appreciation of their services over the last years, which has led to a sort of rediscovery that has been coupled with an interest in developing its potential in terms of cultural tourism. Along the irrigation channels, there always are service pathways for maintaining and repairing the channels. These are normally not open to the public since they pass on private property and the irrigation communities have been hesitant to give public access due to having to cope with the wear and tear of the use of the path with nothing in return. But in the villages mentioned above, the payment-for-service contracts have been able to overcome the resistance of the landowners along the irrigation channels in return for recognition and assistance in maintenance and repair derived from public use. The success of the proposal of developing the service pathways into cultural routes is a consequence of this recognition, as they are areas of high cultural, environmental, and landscape value. Apart from recognizing the ecosystem services that these irrigation communities have always provided, the contracts thus also guarantee public access to a least one of the service pathways along the ditches that can be turned into a coherent cultural route, which is then signposted, homologated by the relevant authorities, and promoted on local, provincial, and specialized sector webpages such as those of hiking associations (Correa Jiménez et al., 2024/in Press; Martín Civantos et al., in Press/2024).

The possibility of developing a cultural tourism offer has thus spurred a contractual development that has become a way to ensure a positive impact on the irrigation communities. For the members of the irrigation communities, in their majority farmers and livestock farmers, the development represents a possibility to diversify their economic activity through the multifunctionality of the traditional agrarian systems and the services and benefits they generate. The alternatives from the agricultural and productive point of view are only intensification and industrialization or abandonment, which both have enormous negative impacts at an environmental and social level and are associated with very short future perspectives from the point of view of environmental sustainability (due to resource depletion) and, therefore, also at an economic level.

These agreements are quite often very simple, particularly because they are village councils that have limited powers and economic capacity. Often the simple fact of receiving the recognition and collaboration of the town council is enough for the irrigation community to see support. Sometimes the town councils can only collaborate with their own means like, for example, some small machinery, construction materials, the wavering of building license fees if needed, administrative support, the lending of municipal premises for the irrigation community to meet, etc. Another element that is often highly valued by both sides is the possibility of collaborating to request aid or subsidies jointly or through the municipality. The money eventually obtained through such channels can then be used wholly or in part by the municipality to work on and invest in the community's irrigation ditches and infrastructure. At issue here are plans and programs to which the irrigation communities do not have access because they cannot be applicants or because they do not have the capacity to request and justify them or to advance the money, such as employment plans or rural, heritage or tourism development plans. The idea of the contracts is that they serve as an umbrella for this stable collaboration and generate a positive dynamic that allows progress, overcoming situations of blockage, working jointly between institutions as something normal which gives greater visibility to the irrigation communities and the irrigation systems.

The agreements should also serve to regulate and resolve problems that tourism may cause like, for example, damage to the irrigation ditch that the irrigation community cannot cope with, increased maintenance needs, or even problems with the theft of agricultural products on farms. The contract is not only a document, but a tool for dialogue between institutions and between neighbours, to solve problems and provide support to the community and the territory.

## 4.5    Innovation and experimental development of contracts

In Figure 4.1, we have outlined how the payment-for-service contracts help alleviating the potential conflict of interests between stakeholders involved in the maintenance and developing of the irrigation system in Granada. Currently, the irrigation communities play a key role in maintaining the historical heritage and providing important community services. Moreover, they are keys for developing cultural tourism by providing public access to service pathways along the historical irrigation systems. However, currently, the irrigation communities do not receive much local acknowledgement of their efforts, neither financially nor socially. Here, payment-for-service contracts between the irrigation communities and the town councils can be a tool for goal alignment and local recognition. As seen in Figure 4.1, the pay-for-service contracts are a tool for bridging the gap between the (1) core resources and activities (access to, and maintenance of, cultural heritage by irrigation communities) and

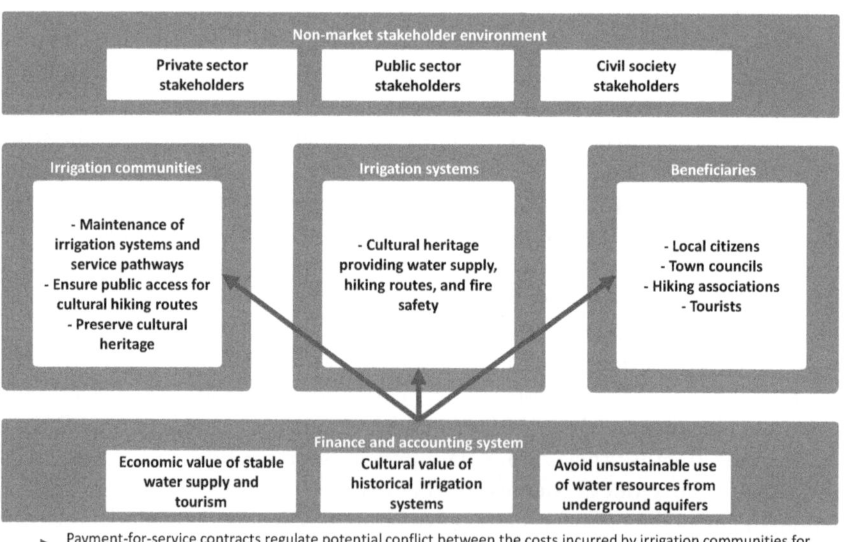


*Figure 4.1* Payment-for-service contracts for historical irrigation systems in Granada.

(2) the needs and wants of customers/beneficiaries (local citizens, town councils, hiking associations, and tourists). Ultimately, the payment-for-service arrangements tie the business model components together and lay the foundation for the development of cultural tourism in the region.

The "payment-for-service" agreements are pioneering and will serve as an example for other irrigation communities and public entities. The main innovation is actually the contracts themselves. As explained above, there is quite a lot of literature on payment for services but mostly from theoretical points of view on the identification and accounting of these services or on the perception of citizens and other stakeholders.[5] However, there are virtually no practical experiences or proposals to put them into practice. In this sense, the contracts that are being developed in Granada are in themselves a novelty. The focus has not been on how many ecosystem services the historical irrigation systems provide or their typology and classification from a theoretical perspective, but rather on negotiating real-life, practical issues directly with the stakeholders involved, which means to develop them experimentally with the irrigation communities and the town councils that represent the closest level of public administration. The main interests of the INCULTUM team were to make a practical proposal, which would intervene directly in and have a practical effect on people's lives and on local economic and productive policies and strategies. To achieve this, the interests and needs of the participating institutions had to be taken into account just as the previously existing relationship between them, including conflicts, which in some cases had been dragging on for a long time. If negotiations had not taken their point of departure in those interests and needs, they would have been much more complicated or have led to failure. Furthermore, the contracts also need to consider the capacities both economic as well as in terms of work force of the two parties and the powers of the town councils, which are actually very limited, since we are talking about small municipalities and their responsibilities do not include agriculture, the environment, nor cultural heritage or water (beyond urban water supply). These are the reasons why the agreements have been based on four elements:

Two of the elements are more generic and have to do with the well-being of the citizens and the territory in general as well as with the good relationship between the municipality and the local irrigation community:

1  Environmental benefits and landscape maintenance.
2  The social (and economic) importance of the irrigation community (all neighbours are also usually community members) and the sense of identity of the community and the irrigation space.

The other two elements are more specific and this is where the issue of the need for the agreement and a more formalized collaboration was raised:

3  Formal recognition of the service of providing urban water supplies, which in many cases depend directly on the irrigation community (recharging

of aquifers and springs, the use of the irrigation ditch to carry water for the drinking water cisterns, transfer of water for supply that historically belonged to the community...).

4 The proposal to make a cultural trail along one of the irrigation channels, as a service that the irrigation community makes available to the municipality, but from which it does not benefit directly.

Added to these main elements, is in many cases the possibility of using the agreement as a way to iron out differences and conflicts, reach agreements to resolve them and put them in writing for mandatory compliance. The intervention of the INCULTUM Granada team has been mediation and facilitation, but also to serve as guarantors of the agreement and its compliance.

The role of the local community is decisive in these often very small municipalities in a rural context, because they influence key aspects such as (1) personal and family relationships; (2) very local identity issues; (3) the relationship of the community or the town council with other local institutions/companies/associations; (4) the relationship with other administrations and institutions outside the territory but with interests or powers that directly influence the local community; (5) regulations on, e.g. environmental or heritage protection, territorial planning, hydraulic planning, agricultural policies; (6) development strategies that depend on regional or central governments and the related public investments, lines of aid, and subsidies and the possibility of accessing them; and (7) the productive context influencing agricultural intensification processes, abandonment processes, or development of rural and cultural tourism.

## 4.6   Perspective impact beyond Granada

The contracts so far formalized and the ones in process are all contracts between a municipality and the local irrigators' community as the only formal stakeholders and signatories. The management of irrigation systems has a very direct influence at the local level and the relationship between the two stakeholders is characterized by relative symmetry in so far as there normally is only one irrigators' community in a municipality which has facilitated the negotiations. This does not mean that other stakeholders are not interested in or have some kind of stake in the process such as local associations or local action groups,[6] but as these stakeholders are not directly involved in the exchange of services at the heart of the contracts, they are not signatories.

In the medium to long term, however, the strategy is to involve more stakeholders, particularly more institutions and more levels of public and semi-public administration, especially those directly related to water management such as the hydrographic confederations, each of which is responsible for the administration of water in a river basin in Spain. But to achieve this will be much more complicated process and require a lot of additional efforts; furthermore, it will almost surely involve changing the type of agreement.

The strategy also is to expand the number of agreements that communities sign with other administrations and institutions within the scope of their powers and interests; it could also be with private entities, for example, entering into the market for carbon credits. However, due to the difficulties that this entails, the focus so far has been on the local level and on formalizing practical agreements and hopefully thereby inspiring further development.

The formula of the agreements is solid and it is replicable in many other areas such as the entire agricultural field where these services and benefits are generated. You only need to change the historical irrigation system for other productive systems and other stakeholders such as extensive livestock farming, the uses of the mountain pastures, forest areas, or traditional pathways. A dialogue must be established between the principal stakeholders and common interests must be identified, both the more generic or abstract as well as the more specific interests focused on the specific services offered and paid for or on the resolution of conflicts or on the mutual benefit of collaborating to raise funds with respect to other administrations. Apart from these general rules, there are no unique formulas that are valid for all situations. Sometimes, other experiences and agreements may serve as models or inspiration, but it is always necessary to adapt it to reality and the local context.

The payment-for-service contracts call for further research in the future. Service-based, financial arrangements are not beyond reproach even though they hold potential for supporting, e.g. local community development and sustainable tourism. For instance, it can be difficult for the parties to define adequate service levels, e.g. when it comes to maintenance of the cultural sites. What is the right service level and how should fulfilment of targets be remunerated? Fundamentally, the value of cultural heritage can be difficult to define, quantify, and include in a contract. Moreover, decisions have to be made about how to measure and control compliance with the payment-for-service contracts. Payment-for-service contracts without functioning control mechanisms will leave room for various types of opportunism from all parties involved, which will undermine the business model. In other words, payment-for-service contracts come with transaction costs, which need to be counted in when deciding to move forward with these financial arrangements. However, the transaction costs are likely to be reduced, as the parties gain more experience with the design and implementation of these contracts.

## 4.7 Concluding remarks

Cultural tourism depends on the successful collaboration between multiple local stakeholders, who work together to keep the cultural sites alive and attractive for visitors. The importance of collaboration is well illustrated in the case of Granada, which shows how the health of the cultural heritage is determined by the joint efforts of town councils, irrigators' communities, hiking associations, and other stakeholders. Ultimately, cultural tourism is a collaborative business model that require concerted action in order to create, deliver, and capture value from visitors. However, some cultural sites are experiencing an uneven distribution of value from cultural tourism, which can potentially undermine the long-term sustainability of cultural attractions. Moreover, lack of coordination and collaboration increase the risk of a sub-optimization and silo-thinking at the expense of the local communities. Here, payment-for-service contracts can help regulating the relationships between the stakeholders and incentivize the parties to work for the benefit of the local community. Moreover, payment-for-service contracts also give formal recognition to the local groups and individuals who play a pivotal role in ensuring access to the cultural sites.

In the future, it will be relevant to explore in more detail how different types of payment-for-service models can be applied to preserve cultural heritage and develop cultural tourism across Europe. It is unlikely that one payment-by-service template works in all contexts. These contractual arrangements need to be adapted to the business model and the stakeholders involved. In the future, it will be relevant for theory and practice to conduct a comparative analysis of contracts used to manage a diverse set of historical and cultural sites. This research can be used to gain experience on the contractual forms considered to be most relevant by the key stakeholders. In the future, it will hopefully be possible to develop a portfolio of easily accessible contractual models, which can be of value for stakeholders promoting cultural tourism at different cultural sites and in different geographical contexts.

## Notes

1 At the moment of writing (Autumn, 2023), a payment-for-services contract has been formalized and signed in the village of Cáñar and another contract has been formalized in Castril, which however still awaits the official signing of the contract. Furthermore, negotiations of other similar agreements are taking place in Jérez del Marquesado, Pórtugos, Bubión, Capileira, Pampaneira, La Tahá, Benamaurel, Galera, and Dílar, all in Granada province.
2 For further information on any of these processes, contact memolab@go.ugr.es.
3 "Arguments in defence of traditional and historical irrigation system", available at https://zenodo.org/record/6523629/files/EN-%20D%C3%ADptico%20 argumentario%20regad%C3%ADos%20hist%C3%B3ricos.pdf?download= 1 (last visited 16 November 2023). For more information, see https://regadio historico.es (last visited 26 November 2023).
4 https://www.euronews.com/green/2023/06/23/spanish-strawberry-growers-deny-using-illegal-irrigation-sparks-controversy (last visited 26 November 2023).
5 One example of such a theoretical contribution is the EFI Policy Brief No. 7 on Payment for Environmental Services (Prokofieva et al., 2012).
6 In fact, in various villages, the Rural Development Group of the Altiplano de Granada has shown an interest in the negotiations and actively worked in favour of the contracts. For information, see altiplanogranada.org (last visited 26 November 2023). Likewise, the Granada INCULTUM-team has established a collaboration with the Association Pasos in relation to various parts of the work that it takes to prepare the contracts. For more information, see https://pasos. coop (last visited 26 November 2023).

## Bibliography

Andersen, M.M. (2023). Barriers to social impact bond implementation: A review of evidence from the UK and US. *International Journal of Public Sector Management*, Ahead-of-print. https://doi.org/10.1108/IJPSM-05-2022-0134

Beulque, R., Micheaux, H., Ntsondé, J., Aggeri, F., & Steuz, C. (2023). Sufficiency-based circular business models: An established retailers' perspective. *Journal of Cleaner Production*, 139431. https://www.researchgate.net/publication/375019249_Sufficiency-based_circular_business_models_An_established_retailers%27_perspective

Bocken, N.M.P., & Geradts, T.H.J. (2020). Barriers and drivers to sustainable business model innovation: Organization design and dynamic capabilities. *Long Range Planning*, 53(4), 101950. https://doi.org/10.1016/j.lrp.2019.101950.

Carter, E., & Ball, N. (2021). Spotlighting shared outcomes for social impact programs that work. *Stanford Social Innovation Review*. https://doi.org/10.48558/84ZM-1Z65

Chesbrough, H. (2007). Business model innovation: It's not just about technology anymore. *Strategy & Leadership*, 35(6), 12–17.

Chesbrough, H. (2010). Business model innovation: Opportunities and barriers. *Long Range Planning*, 43(2–3), 354–363. https://doi.org/10.1016/j.lrp.2009.07.010.

Correa Jiménez, E., Abellán Santisteban, J., Martín Civantos, J.M., Román Punzon, J.M., Bonet García, M.T. (in Press/2024). Senderos culturales a través de acequias históricas en el altiplano granadino: el proyecto INCULTUM. In Sorroche Cuerva and Ruiz Álvarez (editors) *Arquitectura excavada y paisaje cultural*. Madrid: Editorial Dykinson.

Ebrahim, A., Battilana, J., & Mair, J. (2014). The governance of social enterprises: Mission drift and accountability challenges in hybrid organizations. *Research in Organizational Behavior*, 34, 81–100. https://doi.org/10.1016/j.riob.2014.09.001

Gassmann, O., Frankenberger, K., & Csik, M. (2014). *Business Model Navigator*. Upper Saddle River, NJ: FT Press. https://businessmodelnavigator.com/about

Gatelier, E., Ross, D., Phillips, L., & Suquet, J.-B. (2022) A business model innovation methodology for implementing digital interpretation experiences in European cultural heritage attractions. *Journal of Heritage Tourism*, 17(4), 391–408. https://doi.org/10.1080/1743873X.2022.2065920

Geissdoerfer, M., Bocken, N.M.P., & Hultink, E.J. (2016). Design thinking to enhance the sustainable business modelling process. *Journal of Cleaner Production*, 135, 1218–1232. https://doi.org/10.1016/j.jclepro.2016.06.067.

Joyce, A., & Paquin, R.L. (2016). The triple layered business model canvas: A tool to design more sustainable business models. *Journal of Cleaner Production*, 135, 1474–1486,

Kirchherr, J., Piscicelli, L., Bour, R., Kostense-Smit, E., Muller, J., Huibrechtse-Truijens, A., & Hekkert, M. (2018). Barriers to the circular economy: Evidence from the European Union (EU). *Ecological Economics*, 150, 264–272. https://doi.org/10.1016/j.ecolecon.2018.04.028.

La Torre, M., Trotta, A., Chiappini, H., & Rizzello, A. (2019). Business models for sustainable finance: The case study of social impact bonds. *Sustainability*, 11(7), 1887. https://doi.org/10.3390/su11071887

Lüdeke-Freund, F., Gold, S., & Bocken, N.M.P. (2019). A review and typology of circular economy business model patterns. *Journal of Industrial Ecology*, 23, 36–61. https://doi.org/10.1111/jiec.12763

Martín Civantos, J.M., Correa Jiménez, E., Bonet García, M.T. (in Press/2024). Historical water management systems and sustainable cultural tourism. The study case of Altiplano de Granada (Spain). In J.M. Martín Civantos and A. Fresa (editors). *Visiting the Margins: Innovative Cultural Tourism in European Peripheries*. London: Routledge.

Mignon, I., & Bankel, A. (2023). Sustainable business models and innovation strategies to realize them: A review of 87 empirical cases. *Business Strategy and the Environment*, 32(4), 1357–1372. https://doi.org/10.1002/bse.3192

Neesham, C., Dembek, K., & Benkert, J. (2023). Defining value in sustainable business models. *Business & Society*, 62(7), 1378–1419.

OECD. (2016). *Understanding Social Impact Bonds*. OECD Working Paper, Paris, France.   https://www.oecd.org/cfe/leed/UnderstandingSIBsLux-WorkingPaper.pdf (accessed November 10, 2023).

Ostervalder, A., & Pigneur, Y. (2010). *Business Model Generation*. Hobroken: Wiley & Sons.

Osterwalder, A. (2004). *The Business Model Ontology—A Proposition in a Design Science Approach*. PhD Thesis, University of Lausanne, Switzerland.

Pedersen, E.R.G., & Andersen, K.R. (2023). Greenwashing: A business model perspective. *Journal of Business Models*, 11(2), 11–24. https://doi.org/10.54337/jbm.v11i2.7352

Pedersen, E.R.G., Earley, R., & Andersen, K.R. (2019). From singular to plural: Exploring organisational complexities and circular business model design. *Journal of Fashion Marketing and Management*, 23(3), 308–326.

Pedersen, E.R.G., Gwozdz, W., & Hvass, K.K. (2018). Exploring the relationship between business model innovation, corporate sustainability, and organisational values within the fashion industry. *Journal of Business Ethics*, 149, 267–284. https://doi.org/10.1007/s10551-016-3044-7

Pedersen, E.R.G., Lüdeke-Freund, F., Henriques, I., & Seitanidi, M.M. (2020). Toward collaborative cross-sector business models for sustainability. *Published in Business & Society*, 60(5), 1039–1058.

Porter, M.E., & Kaplan, R.S. (2016). How to pay for health care. *Harvard Business Review*. https://hbr.org/2016/07/how-to-pay-for-health-care

Prokofieva, I., Wunder, S., & Vidale, E. (2012). *Pagos por Servicios Ambientales: ¿Una oportunidad para los Bosques Mediterráneos?* EFI Policy Brief #7. Joensuu: European Forest Institute.

Reinhold, S., Zach, F.J., & Krizaj, D. (2017). Business models in tourism: A review and research agenda. *Tourism Review*, 72(4), 462–482. https://doi.org/10.1108/TR-05-2017-0094

Ringvold, K., Saebi, T., & Foss, N. (2023). Developing sustainable business models: A microfoundational perspective. *Organization & Environment*, 36(2), 315–348.

Russo-Spena, T., Tregua, M., D'Auria, A., & Bifulco, F. (2022). A digital business model: an illustrated framework from the cultural heritage business. *International Journal of Entrepreneurial Behavior & Research*, 28(8), 2000–2023. https://doi.org/10.1108/IJEBR-01-2021-0088

Saebi, T., Lien, L., & Foss, N.J. (2017). What drives business model adaptation? the impact of opportunities, threats and strategic orientation. *Long Range Planning*, 50(5), 567–581. https://doi.org/10.1016/j.lrp.2016.06.006.

Savell, L., & Airoldi, M. (2020). Outcomes-based contracts in a time of crisis. *Stanford Social Innovation Review*. https://doi.org/10.48558/A0K2-NS30

Teece, D.T. (2010). Business models, business strategy and innovation. *Long Range Planning*, 43(2), 172–194.
Vermunt, D.A., Negro, S.O., Verweij, P.A., Kuppens, D.V., & Hekkert, M.P. (2019). Exploring barriers to implementing different circular business models. *Journal of Cleaner Production*, 222, 891–902. https://doi.org/10.1016/j.jclepro.2019.03.052.
Weller, A., & Pedersen, E.R.G. (2018). *Pay for Success Literature Review: A PreCare Report*. Copenhagen: Copenhagen Business School, CBS.
Yunus, M., Moingeon, B., & Lehmann-Ortega, L. (2010). Building social business models: Lessons from the Grameen experience. *Long Range Planning*, 43(2–3), 308–325. https://doi.org/10.1016/j.lrp.2009.12.005.

# 5 Navigating landscapes

## Approaches to data collection and analysis in tourism

*Karol Jan Borowiecki, Maja Uhre Pedersen, Sara Beth Mitchell and Shahedul Alam Khan*

### 5.1 Introduction

As we stand at the threshold of a new epoch in cultural tourism, the significance of data analysis in this domain cannot be overstated (Borowiecki et al., 2023a). The advent of data collection and analysis in tourism marks a paradigm shift, transitioning from intuition-based decision-making to insights driven by data.[1] In this chapter, we endeavour to unfold the layers of complexity in cultural and nature-based tourism, addressing the potential challenges and unveiling the latent opportunities within.[2] At the heart of our discourse is the conviction that data, in its myriad forms, is the key to unlocking the secrets of tourist behaviours, cultural influences, and their interplay with the host communities. Data is also essential to understand the impact of any innovative action within the tourism sector. Here, we embark on a quest to navigate the intricacies of data collection, its purpose, its potential, and its profound impact on the realm of tourism. The aim of this chapter is to illustrate the importance of data, and how data collection and the analysis can become an integrated part of project planning and management.

Before starting to collect data, it is important to determine the purpose of the data collection. Cultural and nature-based tourism can be considered as an interplay of history, heritage, nature, and humans that makes it challenging to track the nuances and complexities defining the field. Data collection can be the foundation of our understanding of tourist behaviours, cultural influences, and the impact of tourism on host communities. Furthermore, data is also an important part of evaluating innovative projects within the tourism sector.

To enhance the tourism experience and informed decision-making, we can use data to obtain wider and/or deeper insights, and to disclose patterns. Data on tourism not only develops knowledge for researchers and professionals but also creates value to the very communities at the centre of tourism. The outcome is versatile, ranging from strategies for sustainable tourism and enrichment of tourist experiences to the protection of heritage. Data in this field must capture the essence of the culture and the travel experience. Deciding the methods of data collection is crucial but can be a challenging decision, considering the time and resource constraints. Cultural

DOI: 10.4324/9781003422952-5

indicators, visitor demographics, economic metrics, and more can do the job; however, it is the objectives of the research that will define the scope. The approach to data collection should be a blend of art and science, considering the cultural sensitivity, ethical soundness, and adaptability to circumstances of the destination. Like any research, the analytical techniques should maintain alignment between the research objectives and collected data. Analysis should be able to transform data into knowledge that will illuminate our preliminary research questions. Tourism is a dynamic domain, where progress measurement goes beyond the quantitative milestones. Sustainable tourism can be an ideal mechanism for gauging progress by tracking the preservation of heritage and optimising the impact on the local community. One of the major challenges that remains is the identification of tactics to use the gathered knowledge in the continuous development of the pilots.

This chapter unfolds with a focus on data integration in project planning and management (Section 2), followed by an in-depth look at data collection and handling biases (Section 3). We then explore data analysis through diverse sources like Eurostat and Tripadvisor (Section 4), concluding with insights on the impact of these methods in cultural tourism (Section 5).

## 5.2    Data management

Data can be used during the different phases of implementation of the cultural tourism projects and should therefore be considered an integrated part of the project and project planning. In the initial phases data can be used to explore the opportunities of action, while the collection of data during and after implementation can be used to assess the impact of the pilot action. In all cases, data can come from different sources, based on the needs and intentions of the project.

The importance of collecting data comes from the power of measuring results at all stages of a project (Gudda, 2011). For example, if the results of a project are not measured, it is impossible to tell success from failure. If the success is not measured, it cannot be rewarded and we cannot learn from it. The use of data is an important part of measuring the results.

There is a large literature regarding both management and program planning and evaluation. The field of tourism and specifically cultural tourism is no exception. Several handbooks have been written to aid the implementation of successful interventions within tourism (e.g., Smith and Richards, 2013). However, in such handbooks, there is little to none coverage on the collection of data.

The data planning should be regarded as an integrated part of the project planning. It covers the part of decisions regarding what data to be collected, what methods and tools to use in data collection, and how the data should be analysed subsequently. It is important to consider the framework of the project to ensure that the outcomes are aligned with the objectives of the project. Overall, project planning in cultural tourism can be considered as a four-stage process, as illustrated in Figure 5.1, each of which we will now discuss more in detail. Common challenges such as forms of data, method of

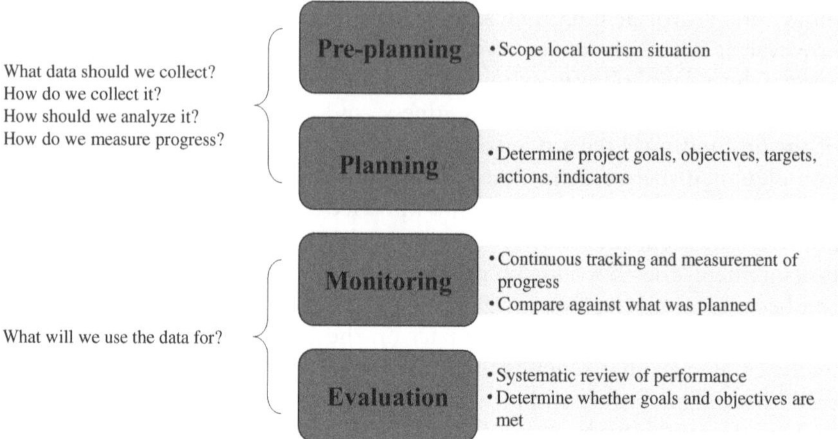


*Figure 5.1* Phases in project planning.

*Notes:* This figure illustrates the different phases of project planning.

*Source:* Own elaboration based on UNECE (2017); IOM (2021).

collection, analytical tools, and measuring progress of the pilot will usually be covered in the first two stages of the data planning. On the other hand, the monitoring and evaluation phases form the purpose of the collection of data (UN, United Nations, 2009).

The pre-planning makes use of existing, to analyse the present condition of the tourism pilot and the surroundings. In this stage, we should ask questions like:

- What is the tourism situation at the pilot site/the wider region like now?
- How does the offering at the pilot site fit into the tourism offering of the wider region?

The second stage is the planning, which includes the determination of goals and objectives for the project. The planning stage can be divided into four distinct stages: (1) formulate goals and objectives; (2) develop strategies; (3) identify polices, programs, projects, and activities; and (4) develop a monitoring and evaluation strategy. The first step is to identify the intended achievements and the time frame for these achievements. Once these have been identified, we can develop the strategies and the action required to reach the goals. Finally, the planning stage should also consider how the action should be monitored and how the project progress should be evaluated. To complete the planning, we should ask questions like:

- What do you want to achieve for tourism in the future and when?
- How will you get from the situation you are at now to where you want to be in the future?

- What specific actions will you take to implement the strategies?
- How will you measure the progress?

In the monitoring stage, we continuously monitor and measure the progress within different areas of the project. Here the progress is compared to the planned metrics such as the maintenance of the time schedule and budget and the attainment of goals. The monitoring also includes identification of causes behind delays and unexpected results together with the adaption of the plans due to changes in circumstances. In this stage, we should ask questions like:

- Are the activities leading to the expected outputs?
- Are activities being implemented on schedule and within the budget?
- What is causing delays or unexpected results?
- Is anything happened that should lead management to modify implementation plan?
- How do stakeholders feel about the pilot?

Finally, the evaluation stage consists of a systematical review of the performance and effectiveness of the goal attainment of the pilot project. This evaluation should be conducted periodically and should also include an evaluation of the resource usage such as time, financial, and human resources. In this phase, the following questions should be asked:

- Have the aims/objectives been achieved?
- How well were resources used?
- Are benefits associated with pilot project likely to last?
- Are pilot project aims/objectives/actions responding to the needs of the local community?
- How well do the project actions fit the needs of the wider region?

In all phases of project planning, it is important to have clear goals. At the initial stages, it should be determined what achievements the project is aiming for, and a clear time frame for both the short, medium, and long terms. This will give stakeholders involved a clear message of all objectives. To set clear objectives, the SMART criteria is a useful tool. SMART is a set of criteria which guides towards setting goals to achieve better results in management and was first proposed in Doran (1981). Even though initially proposed as a tool for management, it has been further developed and is widely cited within the program planning/evaluation literature (see, e.g., Bjerke and Renger, 2017; Chen, 2014; Gudda, 2011). The objectives must be *specific*, where the actions, roles, responsibilities, and accountabilities must be clearly mentioned. They must be *measurable* by developing appropriate metrics to observe, analyse, and verify the outcomes of the efforts. The objectives must be *attainable* considering the given time and resources, and *relevant* considering whether they are useful in obtaining the expected outcomes and results. Finally, the objectives must be *time-bounded*, i.e.

they must assign a time budget to achieve the objectives. Once the objectives have been determined, the actions should be specified together with the targets. This is the stage where data can be linked to the project. In determining the targets, we should start by considering the baseline situation, i.e. the status quo, without the implementation of the new project. To determine the baseline situation, historical data covering the last few years, for example, three to five years, can be used to see the current levels and trends in the sector. The baseline can be used to reflect upon the project objectives, actions, and targets. After having determined the baseline, a few realistic and feasible potential scenarios after implementation can be compared to this. The potential scenarios should include a best-case and a worst-case scenario together with the most realistic. This part of the planning helps detecting the best and worst possible outcomes of the project.

There are some potential pitfalls and barriers to effective project planning. A common challenge is to set too many goals, which will lead to an overwhelming work burden. It is also important to avoid goals that are not well defined or too broad. A poorly defined goal can create confusion and a lack of focus, hampering the effectiveness of the project. Finally, another obstacle is determining unachievable aims considering the given time frame, leading to a potential failure.

A good way to overcome the above challenges is to use the *Problem tree analysis* where the problem is mapped out to problem, causes, and effect. Potentially, the pilot project has the aim of reversing the final effect, e.g. to reverse depopulation in the area. The main problems behind the cause will often require wider policy changes that are beyond the scope of the project. Instead of looking directly at the problems, they can be mapped out into causes towards which the pilot action can be directed.

## 5.3    Data collection

Data can be collected during the different phases explained in Section 5.2 and can come from different sources, primary and secondary. Since the collection of primary data can be both costly and time consuming, it is good practice to investigate what data already exist from other studies. In this chapter, we consider three types of collected data each with its own specific advantages and disadvantages. In the remaining part of this section, the main focus will be on the collection of primary data specifically related to the tourism sector, i.e. visitor surveys. However, part of the discussion can also apply to other data sources that we will present in Section 5.4.

### 5.3.1    *Data in the pre-planning phase*

As a part of the pre-planning process, it is important to develop a profile of the local tourism industry and a profile of the visitors. The key elements of the tourism industry profile are illustrated in Table 5.1 panel (a) together with the list of actions to take. The local tourism industry profile is used to identify the scope of tourism and the related problems to be solved. The first two elements consider existing factors that have an impact on the project and help

*Table 5.1* Content of tourism and visitor profiles

| | *Panel (a) Local tourism industry profile* |
|---|---|
| Tourism resources and assets | List of attractions |
| | List of Facilities |
| Institutional elements | List of stakeholders |
| | Overview of local tourism sector |
| | Local government capacity |
| | Infrastructure |
| Tourist concerns | List of possible concerns of tourists |

| | *Panel (b) Visitor profile* |
|---|---|
| Visitor demographics | Age |
| | Gender |
| | Nationality |
| | Who they travel with (number, type) |
| Characteristics of stay | Length of Stay |
| | Overnight stay |
| | Travel as part of a package tour |
| | Which attractions do they visit |
| | What is the estimated spending |
| Site discovery | How did they hear about the site |
| | What resources were used to explore/learn before visiting |
| Site-level information | Profile visitors from wider area or region |
| | Attract new tourist segment(s) |
| | Draw part of existing tourist segment(s) |

*Notes:* This table illustrates the content of a profile of the local tourism industry in panel a and of a visitor profile in pancl b.

*Source:* Own illustration based on UNECE (2017).

identifying what new possibilities can be created. The third element consider possible concerns of visitors such as lack of public transportation, disability access, language barriers, theft, hazards, etc. These are all areas of improvement of the existing infrastructure and possibilities.

The elements of the visitor profile can be seen in Table 5.1 panel (b), together with a list of content/questions to ask about the visitors. The visitor profile helps identifying the market of local tourists and their needs. The first part of the profile covers basic information about demographics of the visitors and the characteristics of the stay. The second part relates to the visited site. Here, it is especially important to know how they have heard about the place. In case of projects creating new attractions/sites, it might not be possible to have site level information about the visitors. In this case, a visitor profile should be created for the wider area/region and it should be considered what segments of tourists to attract. In some cases, the visitor profile will show that the new site should attract new segments, while, in other cases, it is possible to draw from the existing tourist segment(s) of the wider area.

## 5.4    Data considerations

Before presenting the collected data, there are some important steps to take. First, the data should be cleaned up, to ensure accurate and reliable results during the analysis. Secondly, the missing values should be treated properly. Finally, different potential biases in the data should be considered and discussed.

To clean up the data, we look for errors such as missing and extreme values. It is good practice to double check the data and create a summary table with min, max, and average values to check for abnormalities. If the data set is large, it can be an option to do nothing, especially if the margin of error is small. However, for small data sets, treating the errors is important. The errors can only be corrected if the accurate answer can be confirmed or if the intention is obvious. Otherwise, the solution could be to delete the incorrect observations. If data is deleted, it is important to assure that the mistakes are not systematic and the exclusion criteria should be transparent.

Treating missing values can be a tricky task. First of all, it is important to get an overview of how many missing values exist and for which variables. It is also good practice to get an understanding of why the values are missing. For example, in a survey, it is always suggested to include an option such as "I don't know", "I don't want to answer", "not applicable", etc. Once the missing values have been identified, one possibility is to determine if the missing values can be imputed based on other available data. If imputation is not a possibility, another solution is to exclude the missing values or variables with many missing values from the analysis. In all circumstances, it is important to determine if the data are missing at random or systematically. If data is systematically missing, the implications or potential bias should be considered.

Finally, different biases can distort the results and conclusions from the actual scenario. It is not always possible to treat biases in the data, but a section should be dedicated to a discussion of the potential biases acknowledging their existence and, in case, they are dealt with, how this is done. Biases can exist in all types of data collection methods, but when the data is collected from surveys, there are two main groups of biases which can occur: respondent biases and researcher biases (IOM, 2021)

Respondent biases are biases related to the answers provided in the survey questionnaire. We will go through some of the common potential biases and how they can be treated. Often, respondent biases can be reduced by constructing an adequate questionnaire which reduces the risk of a biased answer. In other cases, the bias should be treated during the initial phases of selecting the representative sample.

**Selection Bias:** Selection bias can arise when certain groups of respondents may systematically agree or disagree to participate. It can be a concern when people volunteer to participate in a study as a respondent, they may answer differently than the people who did not volunteer.

**Non-response bias:** This refers to respondents who refuse to or are not able to respond to the study. In such a case, the collected data will not properly represent the perception of the target population.

- **Attrition bias:** If a study requires more than one round of answers from the same respondents, there is a risk of attrition bias. The data collected may fail to represent the population, if respondents drop out of the study mid-way through and force the project personnel to adjust the sample.

- **Acquiescence bias**: When respondents have the tendency to respond positively towards every question in the survey, it creates biases. In this regard, questions can be revised in a form to get the actual reply from the respondents.
- **Social desirability bias**: In this bias, the respondents tend to give what they think is the socially acceptable answer. To deal with this bias, the question asked should be indirect, so that they do not have the pressure of social acceptance while answering (see e.g. Fisher, 1993).
- **Anchoring bias**: Regarding this bias, the respondent's answer is influenced by a reference point. In answering questions, respondents may rely on the information given in the earlier stages of the survey. This information can work as an anchor and influence them to give a biased answer. The best way to deal with this bias is to avoid leading or suggesting language, provide diverse perspectives, and randomise the order of questions while preparing the survey questionnaire.
- **Recall bias**: In some cases, the respondents may have difficulty remembering certain information. In such issues, we may refer to them some key facts that will help them to recall the relevant information.

Of particular relevance in when conducting visitor surveys is the social desirability bias, which can have important implications for the results. In Dahlgren and Hansen (2015), they explain how the nationality of the interviewer can influence the answers of the respondents. When the interviewer is of the same nationality as the target destination, they show that respondents will assess more positively the attraction. Therefore, the quite common practice of a local or domestic interviewer who interviews tourists at a destination is severely prone to biased results and should be taken into account when planning the survey.

The second category of biases is related to the person conducting the survey/analysis, known as researcher bias.

- **Question-order bias**: The sequence of questions may influence the response and create bias. Selection of words and presentation of ideas may create a partial impression in the mind of the respondents and influence the subsequent answers. To reduce the impact of this bias, general information can be sought before specific information.
- **Leading questions/wording bias**: Wording as mentioned earlier may nudge the respondent to a particular answer. To reduce this bias, researchers should frame questions using the language familiar to the respondents and refrain from paraphrasing their responses from their own perspective.
- **Confirmation bias**: Researchers focus on information that reinforces or confirms their hypothesis or belief. To reduce the bias, the personnel should frequently review the imprints of respondents and question existing assumptions and hypotheses.

### 5.4.1 *Presenting the results*

The presentation of the data is vital in the communication of the findings of the pilot project. The presentation and discussion of the data should be based on knowledge and understanding of the data and the topic of study. Furthermore, the data should be put in the context of short- and long-term trends and explore relationships, causes, and effects. In the following, two common ways to present data will be explained: tables and charts.

Smaller presentation tables are usually used to supplement the information given in the text and contain key figures of the results, i.e. summary statistics. On the other hand, reference tables are longer tables that contain the exact data. These are usually only referred to, and not presented directly in the text.

The charts are used as a way to visualise the results, and a well organised graph can contain large amounts of information. There are different types of charts each with a specific purpose. Deciding which one to use depends on the results we want to illustrate and what knowledge the reader should obtain from the graph. It is also important to consider the target reader and adapt the charts to the level of understanding of the reader. Charts are very good when we want to compare variables, e.g. comparing the number of visitors in two different periods. They can also be used to show changes over times, e.g. a line chart showing the change in the number of visitors over the last years. A chart showing the frequency distribution can illustrate occurrences within different categories such as visitors using different types of transportation. Finally, charts are useful to show correlations between variables such as the correlation between the number of tour guides in a location and the number of visitors.

The following guidelines prepared by UNECE (2009) are useful to take into consideration when presenting the data:

- **The target group**: Tailor the writing according to the knowledge and interests of the target group.
- **The role of the graphic in the overall presentation**: Graphs can only add value to the presentation when aligned with the intended message of the report in terms of highlighting contrasts or emphasising trends.
- **How and where the message will be presented**: The approach to presentation will vary depending on the platforms and the audience.
- **Contextual issues that may distort understanding**: It is important to consider the socio-cultural, historical, or economic issues that may distort the understanding of the audience.
- **Whether textual analysis or a data table is the better solution**: Plain textual analysis can be more effective than a complicated table in highlighting the valuable insight of a circumstance.
- **Accessibility considerations**: Everyone should be able to access and understand the data, regardless of technology and disabilities.
- **Consistency across data visualisations**: The use of colours, scales, and labelling norms should be consistent across data visualisations.

- **Size, duration, and complexity**: Consider that, when a long report is presented to the audience, requiring a huge amount of time to read, it makes the understanding more complex.
- **Possibility of misinterpretation**: When readers lack the literacy to interpret graphs, charts, and complex statistics, it creates possibilities for misinterpretation.

In conclusion, when presenting data, it is important to have a clear message and to keep the information simple without providing unnecessary information. It is also important to make sure that missing values and abbreviations are explained properly in the text. Finally, the target audience should be kept in mind.

### 5.4.2   Visitor surveys

We will here cover some of the important considerations the pilot should include when conducting visitor surveys to obtain data. The survey can be conducted in different ways, most commonly via an online survey or a physical survey (e.g. paper questionnaire). The visitor survey should collect data on demographics of the visitors (what kind of people visit, where are they from, how long are they staying, etc.). This survey can also include questions about the visitor experience (how would they rate their trip, how likely are they to recommend the site to others, etc.). In some cases, the pilot project involves digital platforms (for examples, see Borowiecki et al., 2016). Here, an effort should be made to collect data on digital engagement and (if feasible) conduct a survey of online visitors. It is possible to collect basic information about digital engagement/website visitors using tools such as Google analytics. Appendix A provides a sample visitor survey with an introduction and questions to be asked.

To have reliable and useful results, visitor surveys should be conducted on a day-to-day basis for all visitors at the site. For more remote sites where this is not possible, an effort should be made to conduct a visitor survey twice a year (once during low season and again during high season). During each of these survey periods, an effort should be made to conduct the survey at least once on a weekday (Monday-Thursday) and at least once on a weekend (Saturday or Sunday). Furthermore, it is important that the survey is distributed randomly to all visitors to assure that the results are representative.

Finally, it is important to consider the content of the survey, to assure that it fits to the potential respondents. For example, people walking a trail may not have the patience to conduct a long survey while online users who actively use the digital tool may have better conditions to take a longer survey. It is also important that the questions are impartial, clear, and precise to reduce potential biases and missing values.

## 5.5   Data analysis

Throughout this chapter, we have emphasised the importance of data in all steps of a project, from pre-planning to implementation and evaluation.

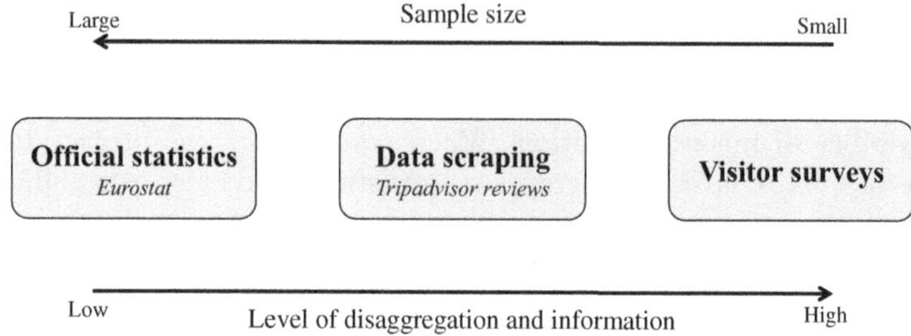


*Figure 5.2* Selected data sources.

*Notes:* This figure illustrates different data sources together with their sample size and level of aggregation and information.

*Source:* Own illustration.

A special focus has been on data collected from primary sources such as visitor surveys. However, the type of data to be collected is specific to the available budget, context, and needs.

The aim of this section is to provide an illustrative example of an analysis of tourism trends in a selected location, using different sources of data, each with its own advantages and limitations. The analysis complements the previous sections, by providing different sources of data, and briefly explaining the advantages and limitations in using each of them. Furthermore, it is also an illustration of how a simple analysis can be conducted, and how the data can be used and presented to show tourism trends. This section also illustrates how alternative and innovative data sources can be implemented. Finally, it improves the understanding of the different data sources, by illustrating what type of information, and at what level of detail, each source can offer. As mentioned in Section 5.3, data can be collected from both primary and secondary sources, and exist at different levels of aggregation and with different levels of information included. We will present data from the following three sources: Eurostat, Tripadvisor, and INCULTUM pilot visitor surveys. Each type of data has its own advantages and disadvantages which should be considered before deciding what data to collect. The list of sources is not exhaustive, but is illustrative of different levels of detail and aggregation. In Figure 5.2 we present an overview of the three levels of data presented in this section. Each of the three selected sources will be explained in the following subsections. As an illustrative example, we concentrate on the Portuguese INCULTUM pilot site and present the data from the three aforementioned sources to show details about tourism and tourists in this location.

### 5.5.1  *Official statistics*

At the highest level of aggregation, we have data from official statistics such as Eurostat. Official statistics have the advantage of being a reliable source and comes with a large sample size and are comparable across countries and over time. However, data is usually highly aggregated both in time and space,

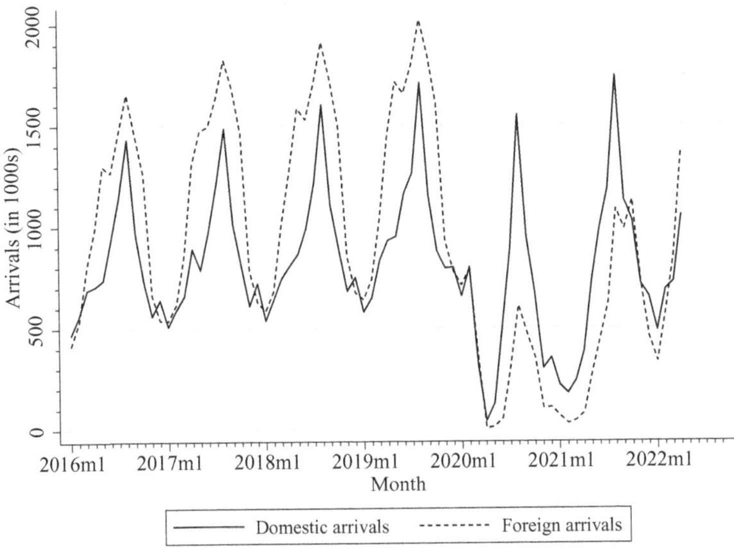

*Figure 5.3* Eurostat domestic and foreign arrivals over time.

*Notes:* This figure shows the number of domestic and foreign Eurostat arrivals over time in Portugal.

*Source:* Official statistics from Eurostat (2023).

meaning that it can be hard to detect effects for smaller units such as the region, city, or attraction level. Usually, official statistics also have a limited amount of information at the individual level, meaning that they cannot be used to establish effects regarding the visitors of a specific location. We present data related to tourism provided by the statistical office of the European Union, Eurostat. Eurostat provides the number of arrivals at tourist accommodations by month and country, and separately for domestic and foreign visitors (Eurostat, 2023). An arrival at a tourist accommodation establishment is defined as a person (tourist) who arrives at a tourist accommodation establishment and checks in. There are made no restrictions on age, meaning that adults as well as children are part of the statistic. Same-day visitors who spend only a few hours (no overnight stay) are excluded from this statistic. In Figure 5.3, we show the number of domestic and foreign arrivals in Portugal over time for the period 2016–2022. From Figure 5.3, there is a clear pattern of seasonality for both domestic and foreign tourists.

### 5.5.2 *Data scraping from Tripadvisor*

At the second level, we have alternative data sources, such as data scraping which can be used to obtain fairly large samples of data and, at the same time, contain more information about the visitors and attractions than the official statistics. In Borowiecki et al. (2024b), we present this new method relying on reviews collected from the travel portal Tripadvisor

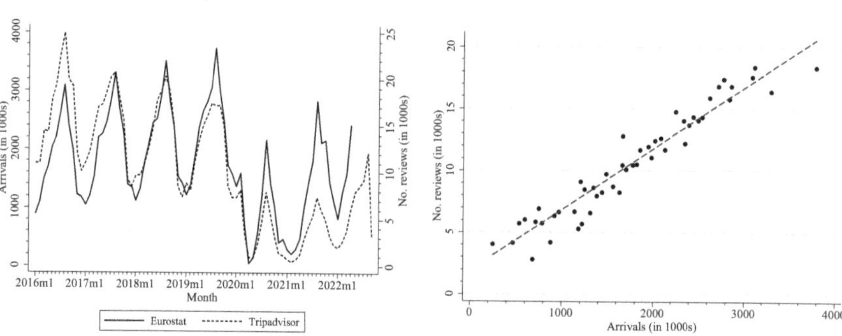

(a)Change in arrival sand reviews over time

(b)Binned scatter plot

*Figure 5.4* Validity tests of Tripadvisor data.

*Notes:* This figure is a visual inspection of the validity of our Tripadvisor data using data from Portugal. Panel (a) shows the change in Tripadvisor reviews over time together with Eurostat arrivals. Panel (b) shows binned scatter plots between number of Eurostat arrivals and Tripadvisor reviews.

*Source:* Arrivals from Eurostat (2023) and Tripadvisor reviews from Borowiecki et al. (2023b, 2024a, 2024b) (see Section 4.2 for details).

for all attractions in various countries. In Borowiecki et al. (2023b, 2024a), we re-apply the method to collect reviews for all attractions in countries in which an INCULTUM pilot site is located. The data contains a list of all reviews in English and native language published by users, together with information about the user posting the review, and all the attractions in each country. This gives us a more detailed set of information, including both the visitors and the attractions. The information included in the data set are date of the review, name and location of attraction, location of user, type of visit (e.g. with family or friends), type of attraction (e.g. museum), rating of the attraction, and travel distance. Before inferring results from an alternative source like Tripadvisor reviews, it is important to validate the data. The validation assures that any results obtained from the data are reliable. Borowiecki et al. (2024b) validate, both through a visual inspection and a formal analysis, the pursued approach by comparing the novel Tripadvisor data to the official statistics from Eurostat. In what follows, we present methods for a visual inspection.

A simple way to validate the data is by comparing the time trends. In panel (a) of Figure 5.4, we show the number of Tripadvisor reviews over time, together with the number of arrivals from Eurostat. The change in the number of reviews over time follows the change in the Eurostat arrivals quite well, indicating that the reviews are a good approximation of tourism flows in Portugal. A second way to inspect the validity is to look at the binned scatter plot in panel (b) of Figure 5.4 where we illustrate the correlation between the two variables, arrivals and reviews. The closer the points are to a straight line, the higher is the correlation. From panel (b) of Figure 5.4, it is clear that they correlate quite well. Given the high correlation, we can establish that the Tripadvisor data is a valid way to analyse tourism flows. Further and more formal validity tests can also be carried out

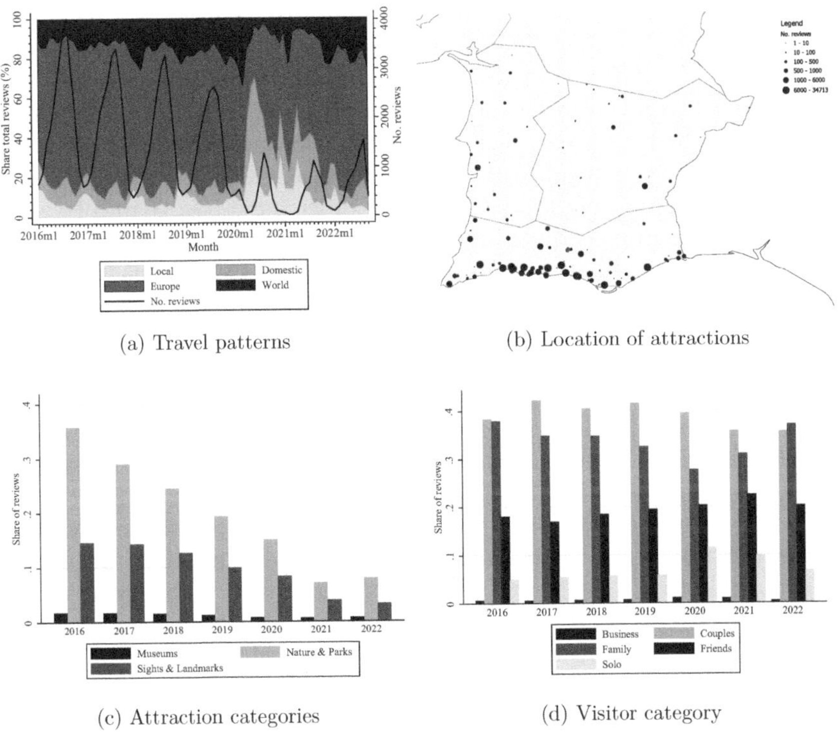

(a) Travel patterns

(b) Location of attractions

(c) Attraction categories

(d) Visitor category

*Figure 5.5* Tourism trends using Tripadvisor data.

*Notes:* This figure shows different results obtained from the Tripadvisor data. Panel (a) illustrates travel patterns for the four travel categories: (1) local, (2) domestic, (3) Europe, and (4) world. Panel (b) shows the location of attractions and number of reviews of each. The grey diamond indicates the approximate location of the INCULTUM pilot site in Portugal. Panel (c) shows the share of reviews of different attraction categories related to cultural and nature-based tourism. Panel (d) shows the share of reviews for different visitor types.

*Source:* Tripadvisor reviews from Borowiecki et al. (2023b, 2024a, 2024b) (see Section 4.2 for details).

by estimating the correlation shown in panel b of Figure 5.4 in a regression setting and looking at the significance of the estimate (see Borowiecki et al., 2023b, 2024a, 2024b).

Once the data is validated, we can use it to look at tourism flows in our selected INCULTUM location in Portugal. Given that we have information about the individual attractions and users, we can look at a smaller unit of observation than the country. This analysis can be used to create a profile of the local tourism trends close to the pilot site and also a broad profile of the visitors in terms of origin and attraction choices. In Figure 5.5, we show different results aggregating our data at the NUTS3 level, and concentrating on the NUTS3 regions close to the location of the Portuguese INCULTUM pilot site. We identify all attractions located within the NUTS3 region where the pilot site is located and attractions in the bordering NUTS3 regions. In panel a of Figure 5.5, we break the total number of reviews for the selected NUTS3 regions into four different travel categories: local, domestic, Europe, and world. The different shades of grey show the share of reviews for each of the four travel categories, out of the total number of reviews. It can be noticed that there is a very high share of European visitors, most of which are from outside Portugal, with the exception of a shorter period following the Covid-19 pandemic in 2020. On the other hand, the share of visitors from outside Europe is quite low. In panel b of Figure 5.5, it is possible to see the location of all attractions in the selected NUTS3 regions. The size of each dot is based on the number of reviews at the attraction, while the red dot indicates the approximate location of the Portuguese INCULTUM pilot site. There is a clear pattern of the location of attractions and the number of reviews, with a higher concentration close to the coast and much less attractions in the inland. In panel c of Figure 5.5, we go more in detail with the type of attractions visited. Tripadvisor categorizes each attraction into one or more categories based on the attraction type. Given that INCULTUM has a special focus on cultural and nature-based tourism, we present results for the three attraction categories most related to these. From panel c of Figure 5.5, it is possible to see the share of reviews in the three categories: museums, nature and parks, and sights and landmarks. In the Portuguese pilot area, there is a high share of visitors to natural sites such as parks, followed by visits to different sights and landmarks. At the same time, there is a downward trend in all three categories, indicating that cultural tourism is trending downwards. Finally, in panel d of Figure 5.5, we illustrate the annual trends in the type of visitors in the five categories: business, couples, family, friends, and solo. The share of visitors, going for business is very low in all years reaching levels well below 5% in all years. The two largest categories are families and couples constituting more than 30% of all visitors each.

### 5.5.3   *Data from visitor surveys*

To complete the presentation of data, we look at the most detailed data, namely the data obtained from visitor surveys conducted at the pilot site. The level of information possible from visitor surveys is very high, while the sample size is usually quite small, given limited resources and time constraints. However, the additional information obtainable from a visitor survey makes this an important part of the evaluation of the pilot action.

An alternative to conducting visitor surveys is to use the surveys conducted by the European Commission. The European commission conducts surveys on travel behaviours and motivations based on a harmonised questionnaire for a large group of countries together with other surveys of national and international travel behaviour. Such surveys are conducted at the trip level and are often of a higher quality than a stand-alone survey. Furthermore, another important feature of such surveys is the stratified, representative sampling procedure and results which can be generalised. For examples of

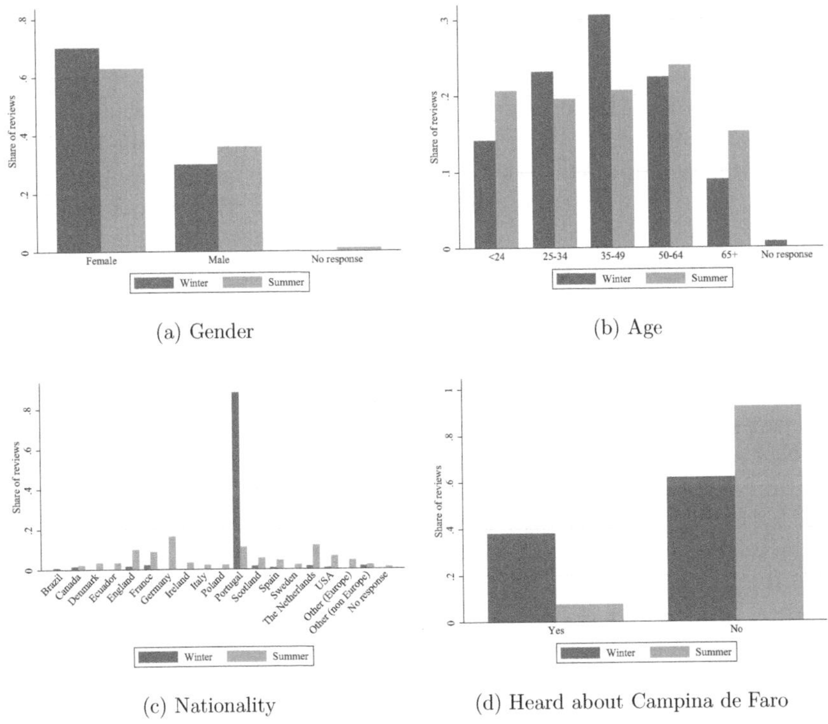

(a) Gender

(b) Age

(c) Nationality

(d) Heard about Campina de Faro

*Figure 5.6* Results from on-site visitor surveys conducted in Campina de Faro (Portugal).

*Notes:* This figure shows results from visitor surveys conducted in Algarve, Portugal during the winter 2022–2023 and the summer 2023. In the winter survey, 134 visitors participated while in the summer survey 92 visitors participated. Panel (a) shows the gender distribution, panel (b) the age distribution, panel (c) the nationality, and panel (d) the answer to the question: have you heard about Campina de Faro?

*Source:* INCULTUM pilot survey data from survey conducted in the Algarve region in Portugal.

the use of such surveys, see Boto-García et al. (2019) who use survey data to analyse the length of stay in a particular location and Vergori and Arima (2020) who study cultural tourism and seasonality.

However, in some cases, there might be very specific requirements for the survey, such as specific questions about the location, which cannot be obtained from more generic surveys. In such cases, a new survey can be conducted to obtain the needed answers.

In Figure 5.6, we illustrate some of the results from two visitor surveys conducted close to the Portuguese INCULTUM pilot site. The first survey was conducted in the period December 2022 to January 2023 and a total of 134 visitors participated in the survey. The survey was mainly targeting residents and visitors in the Algarve region and the aim was to understand the perception of the name "Campina de Faro". The second survey was conducted during the summer 2023 and was targeted mainly towards North-European beach tourists with a total of 92 respondents. The aim of the second survey was similar, with a focus on the perception and knowledge of "Campina de Faro". Given the different target populations of the two surveys, they are not entirely comparable. However, they both have a focus on the knowledge of the pilot site, and hence the results show how different populations perceive the location, which make the comparison useful. For more details about how the surveys were conducted, we refer to Chapter 3, where sampling strategy and representativeness are explained more in detail.

In panel a of Figure 5.6, it is possible to see the gender distribution of the respondents while in panel b we show the age distribution. In both surveys, more than two-thirds of the respondents are women, with a slightly higher share in the survey conducted during the winter. In terms of age distribution, there are some differences between the two surveys. In the winter survey, there is a higher share of visitors between 35 and 49 years of age, while, in the summer survey, respondents are more evenly distributed across all age intervals. In panel c of Figure 5.6, it is possible to see the nationality of the respondents. In the winter survey, there is a high share of Portuguese respondents, more than 80% of the sample. This is expected, since this survey was targeted towards residents. In the summer survey, the distribution of nationality is more diverse. The highest share of respondents is from Germany, followed by the Netherlands and England. Visitors from both Scandinavia and Southern Europe also have good shares. Finally, the share of visitors from Portugal is also significant in the summer survey. In panel d of Figure 5.6, we show the responses to the question "Have you heard about Campina de Faro?" Clearly, a large share of respondents did not know about the place, especially in the summer survey where more than 90% answered "No" against about 60% in the winter survey. This difference is reasonable, given the different target populations, and indicates that especially foreigners are not very aware about the pilot site. This is an important point given the results from panel a of Figure 5.5 showing that a large share of tourists visiting the region in which the pilot is located originate from Europe and outside Portugal.

Together, the analysis of the three selected data types gives a comprehensive overview of tourists and tourism trends in the Algarve region in Portugal.

## 5.6   Conclusion

Some regard data to be as valuable as gold. Others contest instead that data is the new oil. Whether data is shiny or black, the consensus is that it is highly valuable. Consequently, data has become a crucial foundation for business decisions and drives economic activities. In some contrast to this, the cultural heritage sector often does not exploit the full potential of data.

In this chapter, we have underscored the pivotal role of data collection and analysis in enhancing and understanding cultural tourism. As we have seen, data serves as a cornerstone in the realm of cultural tourism, not only for understanding current trends and visitor behaviours but also for planning and implementing successful cultural tourism projects.

In the initial phases of project planning, data equips stakeholders with insights to explore opportunities and set realistic goals. The alignment of data planning with project planning is crucial, ensuring that data collection and analysis are integral to each phase of a project. This holistic approach enables a comprehensive understanding of both the tourism situation at pilot sites and the broader tourism landscape.

The collection of data, whether primary or secondary, presents its own set of challenges and opportunities. Primary data, particularly from visitor surveys, offers invaluable insights into visitor demographics, behaviours, and perceptions. However, it also requires careful consideration in terms of survey design, implementation, and data cleaning processes to mitigate biases and errors.

Secondary data sources vary significantly in their scope and depth. For instance, Eurostat's official tourism statistics provide a broad overview of tourism trends and visitor profiles, offering reliable and large-scale data. However, they often lack granularity and specificity, particularly when it comes to the finer details of individual tourist experiences and behaviours.

On the other hand, novel data-science approaches, such as the analysis of reviews from a leading travel portal, open up new avenues for in-depth and granular insights. Unlike traditional statistical data, Tripadvisor reviews offer a wealth of disaggregated information. This includes detailed feedback on tourist experiences, preferences, and behaviours. More significantly, these reviews can reveal patterns in tourists' past travels, their specific interests in various aspects of cultural sites, and their subjective evaluations of their experiences. Such data can shed novel light on the nuances of visitor engagement and satisfaction, providing a more detailed and nuanced picture of cultural tourism dynamics. However, it is important to note that the collection and analysis of this type of data are neither cheap nor easy, requiring specialised skills and resources.

Data presentation, a critical step in the process, demands careful consideration to ensure clarity, relevance, and accessibility. The use of tables, charts, and other visual aids must align with the intended message and audience, facilitating effective communication of the findings.

The case studies, particularly the Portuguese INCULTUM pilot, illustrate the practical application of data collection and analysis in cultural tourism. These examples highlight the diversity of data sources and methodologies, as well as the depth of insights they can provide into cultural tourism dynamics.

The realm of cultural tourism is on the cusp of a transformative era, propelled by the integration of comprehensive data collection and analysis. This evolution transcends traditional decision-making and project planning, paving the way for a deeper, more nuanced understanding of cultural tourism dynamics. As we navigate this ever-evolving landscape, the strategic harnessing of data emerges not just as a tool, but as a vital catalyst in sculpting sustainable and enriching cultural tourism experiences. Looking ahead, it is this symbiosis of data and cultural insight that promises to redefine the contours of the industry, driving innovation and fostering a more connected and culturally enriched world.

## Notes

1 For an early empirical study of cultural tourism, refer to Borowiecki and Castiglione (2014), who investigate the association between participation in cultural activities and tourism flows in Italian provinces.
2 For a definition of the term cultural tourism see e.g. Du Cros and McKercher (2020) and nature-based tourism Kuenzi and McNeely (2008).

## References


Bjerke, M. B. and Renger, R. (2017). Being smart about writing smart objectives. *Evaluation and Program Planning*, 61:125–127.

Borowiecki, K. J. and Castiglione, C. (2014). Cultural participation and tourism flows: An empirical investigation of Italian provinces. *Tourism Economics*, 20(2):241–262.

Borowiecki, K. J., Forbes, N. and Fresa, A. (2016). *Cultural Heritage in a Changing World*. Cham: Springer.

Borowiecki, K. J., Gray, C. M. and Heilbrun, J. (2023a). *The Economics of Art and Culture*. New York: Cambridge University Press.

Borowiecki, K. J., Pedersen, M. U., Mitchell, S. B. and Khan, S. A. (2023b). *Final Findings Analysis Report INCULTUM*. Deliverable D3.3, Visiting the Margins. INnovative CULtural ToUrisM in European peripheries.

Borowiecki, K. J., Pedersen, M. U. and Palomeque, M. (2024a). Putting the Periphery on the Map: Tourism activity measured with big data. Working Paper.

Borowiecki, K. J., Pedersen, M. U. and Mitchell, S. B. (2024b). *Using Big Data to Measure Cultural Tourism in Europe with Unprecedented Precision*. Discussion Papers on Economics, Working Paper No 5/2023, University of Southern Denmark. https://journals.sagepub.com/doi/10.1177/13548166241273604?icid=int.sj-abstract.citing-articles.1

Boto-García, D., Baños-Pino, J. F. and Álvarez, A. (2019). Determinants of tourists' length of stay: A hurdle count data approach. *Journal of Travel Research*, 58(6):977–994.

Chen, H. T. (2014). *Practical Program Evaluation: Theory-Driven Evaluation and the Integrated Evaluation Perspective*, 2nd Edition. Thousand Oaks, CA: SAGE Publications.

Dahlgren, G. H. and Hansen, H. (2015). I'd rather be nice than honest: An experimental examination of social desirability bias in tourism surveys. *Journal of Vacation Marketing*, 21(4):318–325.

Doran, G. T. (1981). There's a S.M.A.R.T. way to write management's goals and objectives. *Management Review*, 70(11):35–36.

Du Cros, H. and McKercher, B. (2020). *Cultural Tourism*. London: Routledge.

Eurostat. (2023). *Eurostat Tourism Statistics*. https://ec.europa.eu/eurostat/web/tourism. Last data update: 05/01/2023.

Fisher, R. J. (1993). Social desirability bias and the validity of indirect questioning. *Journal of Consumer Research*, 20(2):303–315.

Gudda, P. (2011). *A Guide to Project Monitoring Evaluation*. Bloomington: AuthorHouse.

IOM. (2021). Methodologies for data collection and analysis for monitoring and evaluation. In *IOM Monitoring and Evaluation Guidelines*. Chapter 4.

Kuenzi, C. and McNeely, J. (2008). Nature-based tourism. In Renn, O. and Walker, K. D. (editors) *Global Risk Governance: Concept and Practice Using the IRGC Framework*, pp. 155–178. Netherlands, Dordrecht: Springer.

Smith, M. and Richards, G. (2013). *The Routledge Handbook of Cultural Tourism*. Routledge.

UNECE. (2009). *"Making Data Meaningful" Guide Series - Parts 1, 2 and 3*. United Nations: Economic Commission for Europe - UNECE.

UNECE. (2017). *Tourism Guidebook for Local Government Units*. https://elibrary.bmb.gov.ph/elibrary/books/tourism-guidebook-for-local-government-units/

UN, United Nations. (2009). *Handbook on Planning, Monitoring and Evaluating for Development Results*. United Nations Development Programme. https://www.undp.org/turkiye/publications/undp-handbook-planning-monitoring-and-evaluating-development-results

Vergori, A. S. and Arima, S. (2020). Cultural and non-cultural tourism: Evidence from Italian experience. *Tourism Management*, 78:104058.

# An Appendix
## Sample visitor survey

**Introduction**

This survey is being conducted by [the surveyor (add name)]. The survey is part of INCULTUM (2021-2024), a HORIZON2020-funded project. The main goal of this survey is to better understand visitors to [add name] and how we can improve the visitor experience. If you agree to participate, we will ask you a set of questions about you and your experiences at [add name]. The survey will take approx. 15 minutes time to be answered. Your participation is voluntary, and all information will be anonymised and kept strictly confidential in accordance with the data protection laws and guidelines.

**Section A.  Visitor demographics**

A.1 What is your gender?

- ☐ Male
- ☐ Female
- ☐ Other
- ☐ Prefer not to respond

A.2 What is your age?

_____________

A.3 What is your country of residence?

_____________

A.4 If you live in [add name], please indicate the county/département/provincia where youlive:

_____________

A.5 What is your marital status?

- ☐ Single
- ☐ Married
- ☐ Widowed

☐ Divorced/separated
☐ Prefer not to respond

A.6 How many dependent children do you have?

_____________

A.7 Which category best describes you?

☐ In full-time employment D In part-time employment D Student
☐ Unemployed
☐ Retired/Pensioner
☐ Housewife/househusband
☐ Other (please specify): _________
☐ Prefer not to respond

A.8 What is the highest level of education you have attained?

☐ Completed secondary school or less
☐ Bachelor's degree or equivalent
☐ Master's degree/PhD or equivalent

**Section B.  Details of visit**

B.1 Have you visited [add name] before?

☐ Yes
☐ No

B.2 If yes, when did you last visit [add name]?

☐ Within last month
☐ Within last year

B.3 What is the main purpose of your visit to this area?

☐ Vacation/holiday
☐ Visiting friends/relatives
☐ Education/training
☐ Conference/large meeting
☐ Business/small meeting
☐ Event
☐ Other

B.4 Which best describes the group you are traveling with?

☐ I am traveling alone
☐ A couple

- [ ] A family with children
- [ ] A group of friends
- [ ] A school group
- [ ] An organised tour group (not school-related)
- [ ] Other

B.5 How did you arrive at [add name] today?

- [ ] Private car, van, or motorcycle (e.g., own, friends, family)
- [ ] Rented car, van or motorcycle
- [ ] Taxi
- [ ] Public bus or coach D Private bus or coach D Train
- [ ] Bicycle
- [ ] Walk

B.6 Which of the following best describes your visit to the area?

- [ ] Day trip
- [ ] Overnight stay

B.7 If you are staying overnight, which city are you staying in?

___________

B.8 If you are staying overnight, how many nights are you staying?

___________

B.9 Which of the following best describes the type of accommodation you are staying in?

- [ ] Hotel, motel, hostel
- [ ] Guesthouse, bed, and breakfast D Short-term rental (e.g., Airbnb) D Caravan, camping
- [ ] Home of friend or relative
- [ ] Second home
- [ ] Other (please describe): ________

B.10 Have you visited any of the following sites in the area? Select all that apply.

- [ ] [add location 1]
- [ ] [add location 2]
- [ ] [add location 3]
- [ ] [add location 4]

**Section C.  Visitor experience**

C.1 How did you find out about [add name]?

- ☐ Friends/relatives
- ☐ Tourist information centre
- ☐ Newspaper or magazine
- ☐ Search engine (do not remember which websites) Travel review site (e.g., Tripadvisor, Google Places) Facebook, blog, other social media
- ☐ [add website]
- ☐ Other (please specify): __________

C.2 What factors were important for you when choosing to visit [add name]? Select all that apply.

- ☐ Quality of experience
- ☐ Good value for money
- ☐ Historic interest
- ☐ Scenery and countryside
- ☐ Peace and quiet
- ☐ Friendliness and hospitality of locals
- ☐ Environmental impact
- ☐ Geographic proximity – I live nearby/I am staying nearby
- ☐ Cultural proximity – I identify with what the site represents
- ☐ By chance – I was just passing by/I was already visiting an area nearby A particular event (please specify):
- ☐ Other (please specify): ____________

C.3 Please rate your visit to [add name] on a scale of 1 (Very poor) to 10 (Excellent).

__________

C.4 How likely are you to recommend [add name] to someone else on a scale from 1 (Very poor) to 10 (Excellent)?

# 6 Participatory models and approaches in sustainable cultural tourism

*Kamila Borseková and Katarína Vitálišová*

## 6.1   Introduction

Participatory models and approaches in tourism have gained widespread acceptance as a cornerstone for sustainable tourism. These models not only assist decision-makers in preserving traditional lifestyles and upholding community values but also play a pivotal role in enhancing the image and brand of a tourism destination. By offering superior customer services and fostering innovation, they strengthen the competitiveness of the destination. A shift from top-down decision-making, participatory models aim to equitably distribute power among all stakeholders, fostering a collaborative environment conducive to mutual benefits in tourism development (Ozcevik et al., 2010; Wang & Fesenmaier, 2007; Cater, 1994; Murphy, 1985; Arnstein, 1969). These participatory approaches align seamlessly with the Agenda 2030 and Sustainable Development Goals, particularly Goals 8, 11, 12, and 14, which emphasize inclusive growth, sustainable communities, responsible consumption, and marine conservation. Robson and Robson (1996) posited that stakeholder participation in tourism offers a framework for achieving sustainable tourism development. This balance of power, as highlighted by Vijayanand (2013), bridges the gap between traditional powerholders, such as governments and investors, and the host communities directly impacted by tourism initiatives. When this equilibrium is achieved, the result is a more inclusive, equitable, and ultimately sustainable tourism development. Tourism, as a product, is an amalgamation of diverse activities, with culture being paramount. This chapter delves deep into identifying and analysing participatory approaches within culture, cultural tourism, and sustainable cultural tourism. The underlying rationale is to accentuate the positive impacts of these models while mitigating the potential adverse effects on social ties, local heritage, and landscape conservation. Consequently, the primary objective of this chapter is to furnish readers with a thorough understanding of participatory models and approaches in culture and cultural tourism, culminating in the introduction of an innovative framework for sustainable cultural tourism.

DOI: 10.4324/9781003422952-6

## 6.2    Cultural tourism and sustainable cultural tourism

In recent decades, the concept of cultural tourism has gained prominence, driven by a surge in international tourists visiting major cultural sites and attractions (Richards, 2018). Culture, as a key element of tourism appeal, facilitates access to heritage, art, creativity, and various cultural activities and practices (Matteucci & Von Zumbusch, 2020). Globalization has prompted many destinations to recognize the significance of culture in enhancing tourism offerings, ensuring authenticity, and bolstering a destination's global appeal. Consequently, niches such as creative tourism, arts tourism, film tourism, and literary tourism have emerged (Smith, 2016). While these niches offer development and marketing prospects, they also present challenges.

Cultural tourism emphasizes the cultural facets of a destination, including its heritage, landscapes, and offerings, often driving tourists' destination choices (European Commission, 2019). It engages travellers with the lifestyle, history, art, architecture, religion, and other cultural elements of a region (Slocum, Aidoo, & McMahon, 2020). Tourism, in this context, is viewed positively, as it can monetize heritage attributes, fostering conservation, community education, and policy influence. When adeptly managed, cultural tourism can spur sustainable local development, benefiting host communities and motivating them to preserve their heritage and cultural practices. It accentuates intangible heritage elements, offering avenues to promote local traditions or historical narratives (Slocum, Aidoo, & McMahon, 2020). A notable subset of cultural tourism is heritage tourism, which is rooted in local landscapes, architecture, traditions, and stories, emphasizing the uniqueness of a place (National Trust for Historical Preservation, 2001). Managing this form of tourism requires a multidisciplinary approach, balancing cultural preservation with revenue generation (Ponna & Oka Prasiasa, 2011).

Cultural tourism, while acting as a catalyst for sociocultural transformation, has its own set of challenges. It plays a pivotal role in stimulating local development, fortifying communities, creating job opportunities, and fostering capacity building. Yet, the very essence of cultural tourism, which often hinges on the allure of authenticity, can inadvertently lead to the commodification and commercialization of cultural heritage. Tourists, driven by their desire for genuine experiences, can sometimes be complicit in this commodification, turning cultural assets into marketable products and diluting their authenticity (Bitušíková, 2021; Smith, 2009).

As a response to these challenges, the concept of sustainable cultural tourism has emerged. This approach emphasizes the integrated management of cultural heritage, tourism activities, and community involvement, striving for a balance that benefits all stakeholders (EC, 2019; McKercher & du Cros, 2002). By placing cultural heritage and its communities at the forefront of decision-making, sustainable cultural tourism ensures that heritage sites are both celebrated and preserved. It champions authentic interpretation and boosts local economies in a sustainable manner. Yet, even this form of tourism is not without its challenges. Many European destinations, popular for their rich cultural offerings, face issues of overuse and overcapacity. This over-tourism not only strains local resources but can also lead to environmental degradation and foster negative sentiments among local communities (EC, 2019; Koens, Postma, & Papp, 2018). Thus, for cultural tourism to sustainable, it is imperative that policies and practices respect and uphold the integrity of cultural heritage, ensuring that its inherent values are not compromised in the pursuit of economic gains (Council of Europe, 2005; Timothy & Nyaupane, 2009). Good example is the project 'Participatory Conversion of Historical Irrigation Systems into Cultural Routes', located in Spain's Altiplano de Granada. This project is dedicated to empowering local communities by leveraging cultural heritage as a catalyst for transformative change. It aims to establish a platform for inclusive decision-making, leading to sustainable development and the preservation of rural heritage. This approach not only respects and revitalizes historical practices but also aligns them with contemporary needs and aspirations, ensuring a balanced and sustainable future for the region (for more information, see Civantos et al., 2023).

Emerging sustainable cultural tourism paradigms are increasingly focusing on strategic planning, networking, and innovative concepts that cater to the evolving preferences of today's travellers. Concepts such as 'slow' tourism advocate for a more immersive and relaxed travel pace, allowing tourists to deeply engage with their surroundings and truly experience a destination (Hall, 2017). 'Authenticity' has become a sought-after commodity in tourism, with travellers seeking genuine experiences that reflect the true essence of a place (Wang, 1999). 'Storytelling' enhances the visitor experience by weaving narratives around cultural and historical sites, making them more relatable and memorable (Mossberg, 2008). The emphasis on 'well-being' reflects the growing trend of wellness tourism, where cultural experiences are intertwined with health and rejuvenation (Lourens, 2007). Furthermore, 'contact with locals' provides tourists with firsthand insights into local cultures, traditions, and lifestyles, fostering cross-cultural understanding and exchanges (Salazar, 2012).

Society's intricate relationship with cultural heritage goes beyond passive observation. It encompasses active engagement in its expression, conservation, interpretation, and utilization. Heritage communities and individuals play pivotal roles in shaping and preserving cultural narratives. Far from being passive recipients or mere 'audiences', they are active participants, contributing significantly to participatory governance, scientific endeavours, and the holistic management of cultural heritage, as highlighted by EC (2019) and Bortolotto (2007). An interesting example of this active participation is the 'Bulliot, Bibracte et moi' project, which focuses on transcribing and digitizing the handwritten excavation notebooks of Jacques-Gabriel Bulliot. This project uniquely involved the public in the transcription process through

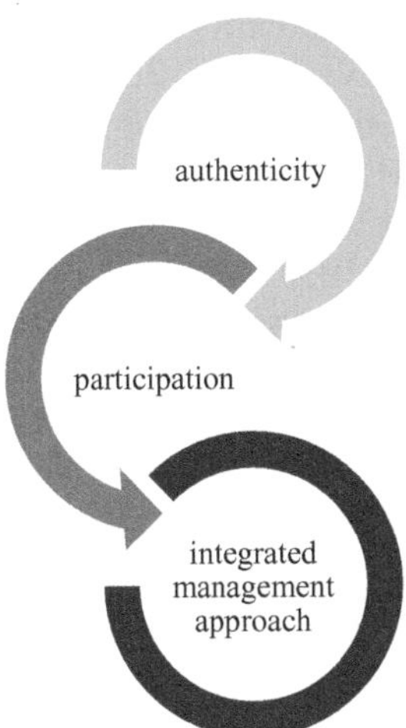


*Figure 6.1* Main features of sustainable cultural tourism.

Transkribus, an online platform where participants trained an AI system to recognize Bulliot's handwriting. What sets this project apart is its empowerment of amateurs to undertake crucial tasks, thereby fostering a sense of community involvement in preserving and interpreting cultural heritage. The project exemplifies how participatory science can unlock the potential of archival materials, engage the public in scientific endeavours, and promote a deeper understanding of archaeological heritage. By combining the expertise of researchers with public contributions, the 'Bulliot, Bibracte et moi' project not only advances archaeological knowledge but also serves as a valuable model for future participatory projects in archaeology, culture, and related fields. For more information on this innovative approach, see Borsekova et al., 2023. Involvement and active building of heritage communities ensures that cultural heritage remains dynamic, relevant, and resonant, reflecting the evolving values and aspirations of the community while preserving its essence for future generations (Smith, 2006).

Figure 6.1 summarizes the main features of sustainable cultural tourism.

Sustainable cultural tourism underscores the importance of an integrated approach, harmonizing cultural heritage, tourism activities, and community involvement (EC, 2019; McKercher & du Cros, 2002). This approach not only ensures that heritage sites are celebrated but also emphasizes their preservation. A key tenet is a promotion of authentic interpretation, which not only resonates with tourists but also support local economies in a sustainable manner. However, this form of tourism is not without its challenges. Over-tourism, for instance, has emerged as a significant concern, straining local resources and potentially leading to environmental degradation (EC, 2019; Koens et al., 2018). To navigate these challenges, it's imperative that policies and practices are crafted with a deep respect for cultural heritage, ensuring its integrity remains uncompromised (Council of Europe, 2005; Timothy & Nyaupane, 2009). As the landscape of tourism evolves, sustainable cultural tourism is increasingly focusing on strategic planning, networking, and embracing innovative concepts. These include 'slow' tourism, 'authenticity', 'storytelling', 'well-being', and fostering deeper 'contact with locals' (Hall, 2017; Wang, 1999; Mossberg, 2008; Lourens, 2007; Salazar, 2012). Cultural tourism, deeply rooted in the rich tapestry of local heritage and traditions, finds its true sustainable potential when participatory approaches are embraced, ensuring that the voices of local communities and stakeholders shape the narrative. Hence, the next section is devoted to outlining participatory approaches in culture, cultural tourism, and sustainable cultural tourism.6.3 Participatory approaches in culture, cultural tourism, and sustainable cultural tourism

## 6.3   Participatory approaches in culture, cultural tourism, and sustainable cultural tourism

Participatory approaches have emerged as a transformative paradigm in the realms of culture, cultural tourism, and sustainable cultural tourism. Rooted in the principle of inclusive decision-making, these approaches prioritize the active involvement of local communities, visitors, and other stakeholders, in shaping cultural narratives and experiences. In the context of cultural tourism, this means co-creating tourism products that authentically represent local heritage, traditions, and values. As the tourism industry grapples with the challenges of sustainability, participatory methods offer a path forward, ensuring that cultural tourism not only celebrates and preserves cultural heritage but also promotes socio-economic benefits for local communities, ensuring a harmonious and sustainable future.

### 6.3.1   *Participatory approaches in culture*

Culture and cultural heritage are pivotal development factors, enhancing the quality of life both within communities and in broader contexts. Local culture plays a crucial role in regional development (Bole et al., 2013; Nared et al., 2013; Nared & Bole, 2020). Echoing this sentiment, the European Commission (2014) emphasized the profound economic and social impacts of cultural heritage and activities, extending beyond cultural tourism to include the promotion of cultural and creative industries. The term 'culture-based development' has gained traction in locales seeking innovative development strategies (Tubadji, 2012). Such development hinges on local actors and their interrelations (Bole et al., 2013). To be a catalyst for development, culture must be effectively evaluated, negotiated, and implemented by a diverse group of stakeholders. The Convention (1972) advocates for the integration of culture into community life, necessitating continuous stakeholder interactions, underscoring the significance of participatory processes (Nared et al., 2013).

These processes should be grassroots, addressing tangible issues (Alfarè & Nared, 2014; Nared, 2014; Nared & Bole, 2020). It is evident in practice by the project 'Multaka: Museum as Meeting Point – Refugees as Guides in Berlin Museums' project. This commendable initiative trains Syrian and Iraqi refugees to become museum guides, enabling them to conduct tours in Arabic for other Arabic-speaking refugees. Not only are these tours offered free of charge, but the project, aptly named 'Multaka' (Arabic for 'meeting point'), also serves as a platform for exchanging diverse cultural and historical experiences (for more information, see Borsekova et al., 2023).

The governance of culture has evolved since the 1980s, with Culture 3.0 marking a significant shift. This paradigm is characterized by innovations stemming from a transformation in cultural production. Technologies like radio, television, and cinema democratized access to cultural content (Sacco et al., 2013, 2018). Culture 3.0 heralded an era where producers proliferated (Potts et al., 2008), enabling individuals to co-design, co-create, and co-produce cultural services (Ciolfi et al., 2008). This dynamic aligns with the concept of 'prosumerism' (Duncum, 2011), where individuals both produce and consume cultural content (UNESCO, 2009).

Central to Culture 3.0 is an active cultural participation, which transcends passive consumption, prompting individuals to harness their skills in the creative process, thereby redefining their social identities (Sacco et al., 2018). Cultural participation encompasses both formal events and informal community activities, reflecting traditions and beliefs (UNESCO, 2009). It can be approached horizontally, promoting participation in specific activities, or democratically, emphasizing citizen influence and control (Eriksson, 2020). Cultural economics has elucidated the complexities of cultural participation, exploring its determinants and its relationship with local and regional development (Ateca-Amestoy, 2008; Ateca-Amestoy & Prieto-Rodriguez, 2013; Falk & Katz-Gerro, 2016). Cultural participation is important for redistributing power, ensuring inclusive future processes. The digital revolution has ushered in Culture 4.0, intertwining culture and technology. This phase recognizes the transformative potential of technologies like artificial intelligence and virtual reality in cultural practices.

### 6.3.2 *Participatory approaches in cultural tourism and sustainable cultural tourism*

Community participation in cultural tourism is a multifaceted concept, intricately weaving the nuances of culture with the dynamics of tourism. While the models of cultural participation have been previously delineated, it's imperative to delve deeper into the role of community engagement in tourism. This section aims to elucidate the significance of community participation, its specificities, and, subsequently, the governance and models in cultural tourism as an alternative form rooted in cultural and tourism development.

Murphy (1985) underscored the significance of the host community in tourism. His work aimed to assess the capacity of local communities to accommodate tourism by identifying their aspirations and objectives. Adopting an ecosystem approach, Murphy emphasized that planning must be intricate, reaching down to the community level. A growing consensus suggests that community participation is indispensable for tourism development (Cole, 2006; Botes & van Rensburg, 2000). Such participation ensures community support for development plans, aligns benefits with local needs, and fosters democratic processes (Tosun & Timothy, 2003). A participatory approach in tourism seeks to deviate from unilateral top-down decision-making. It aims to distribute power equitably among stakeholders, fostering a collaborative environment (Ozcevik et al., 2010; Arnstein, 1969). Haywood (1988) defines participation as a shared decision-making process involving all stakeholders. This approach values the preservation of traditional lifestyles and community values (Murphy, 1985; Wild, 1994; Cater, 1994). Collaborative tourism emphasizes shared experiences, pooling resources, and collective problem-solving (Vernon et al., 2005). Wang and Fesenmaier (2007) highlight its role in enhancing destination branding, product development, and fostering innovation. Empowering communities is essential, recognizing them as integral to the cultural product and addressing their concerns (Timothy, 2011).

Sustainable tourism acknowledges the pivotal role of community engagement (Cole, 2006). Byrd (2007) identifies four stakeholder groups in sustainable tourism: present tourists, present host communities, future tourists, and future host communities. The success of sustainable tourism hinges on the support of these stakeholders (Gunn, 1994). Robson and Robson (1996) advocate for a balanced power dynamic between traditional power holders and host communities, ensuring equitable and sustainable tourism development.

Community-based tourism (CBT) emerges as an alternative, focusing on the benefits for residents in developing regions. It promotes cultural interactions, hospitality services, and biodiversity conservation (Kiss, 2004; Luccetti & Font, 2013). CBT emphasizes the importance of local control, shifting the reins of tourism from external entities to the community itself (Simpson, 2008; González-Herrera et al., 2022). The World Tourism Organization recognizes CBT's potential to preserve culture, foster innovation, and provide educational opportunities (WTO & UNEP, 2005).

In the realm of sustainable cultural tourism, participatory approaches have emerged as a basis for ensuring both the preservation of cultural assets and the equitable distribution of tourism benefits. As Smith and Richards (2013) noted in their seminal work, the active involvement of local communities and other stakeholders in decision-making processes not only engenders a sense of ownership but also ensures that tourism strategies are congruent with local values and aspirations. The transformative potential of information and communication technologies (ICTs) in this context cannot be overstated. Johnson and McCarthy's (2019) study underscores how ICT can be

| Cultural Participation | Participation in Cultural Tourism | Participation in Sustainable Cultural Tourism |
|---|---|---|
| •Inclusive decision-making.<br>•Active involvement of local communities, stakeholders, and visitors (formal and informal).<br>•Co-creation, co-design, and co production of cultural services.<br>•Active cultural participation.<br>•Can be approached horizontally (specific activities) or democratically (citizen influence and control).<br>•Cultural participation is pivotal for redistributing power and ensuring inclusive future processes.<br>•Digital revolution (Culture 4.0) emphasizes active engagement facilitated by technologies.<br>•Bolsters social inclusion, fosters entrepreneurship, and addresses societal challenges.<br>•Recognizes the transformative potential of technologies in cultural practices. | •Co-creation of tourism products that authentically represent local heritage, traditions, and values.<br>•The significance of the host community in tourism.<br>•Aims to distribute power equitably among stakeholders.<br>•Perserves the traditional lifestyles and community values.<br>•Collaborative tourism emphasizes shared experiences, pooling resources, and collective problem-solving.<br>•Empowers communities, recognizing them as integral to the cultural product.<br>•Community-based tourism (CBT) focuses on benefits for residents, promoting cultural interactions, hospitality services, and biodiversity conservation. | •Ensures that cultural tourism celebrates and preserves cultural heritage.<br>•Promotes socioeconomic benefits for local communities.<br>•Recognizes the central role of community engagement.<br>•Advocates for a balanced power dynamic between traditional power holders and host communities.<br>•Emphasizes the importance of local control in tourism.<br>•Active involvement of local communities and stakeholders in decision-making processes engenders a sense of ownership<br>•ICT can be leveraged to both preserve and dynamically showcase intangible cultural heritage.<br>•Championing participatory approaches underpinned by traditional community engagement mechanisms and technological innovations. |

*Figure 6.2* Characteristics of cultural participation, participation in cultural tourism, and participation in sustainable cultural tourism.

leveraged to both preserve and dynamically showcase intangible cultural heritage, thereby enriching the visitor experience. The efficacy of Participatory Action Research in fostering community-centric initiatives that drive sustainable change has been highlighted by Thompson et al. (2018). In this digital age, the nexus of culture, tourism, and technology offers a promising trajectory for sustainable development (Rodriguez & Moretti, 2020).

Figure 6.2 presents key characteristics of cultural participation, participation in cultural tourism, and participation in sustainable cultural tourism. These characteristics highlight the importance of active involvement, co-creation, and the integration of technology in shaping cultural narratives, tourism products, and sustainable practices. For cultural tourism to realize its full potential in terms of sustainability, it is imperative to support participatory approaches, underpinned by both traditional community engagement mechanisms and cutting-edge technological innovations including digitalization. Therefore, the next section is devoted to digitalization and sustainable cultural tourism.

## 6.4    Digitalization and sustainable cultural tourism

In Culture 4.0, traditional cultural expressions are reimagined through digital platforms, blurring the lines between creators and users. It emphasizes active engagement, facilitated by digital technologies, allowing individuals to shape the cultural landscape in novel ways. Cultural participation has profound social and economic implications as it can empower social inclusion, foster entrepreneurship, and address societal challenges. High levels of cultural participation can foster support for investment in the cultural sector (OECD, 2021).

Digital transformation is revolutionizing every sector, with cultural tourism being no exception. Sonkoly and Vahtikari (2018) argue that digitalization democratizes cultural heritage, making it more accessible to the masses. This transformation, as defined by Mergel et al. (2019) and Margiono (2021), leverages technology to enhance governance, creating value for both consumers and businesses. Vial (2019) emphasizes the role of information, computing, communication, and connectivity technologies in this shift. The intricate relationship between digital technology, culture, and tourism has been explored by numerous scholars (see, e.g. Cipolla et al., 2011; Logan et al., 2015). The European Commission (2019) notes that digitalization encompasses economic, social, cultural, and organizational transformations, all driven by digital technologies.

Seifert and Rössel (2022) introduce the concept of digital participation, which they argue is a step beyond mere connectivity. It's about how individuals immerse themselves in the digital realm, interacting with a myriad of online services and content. This sentiment is echoed by Stratigeaand Katsoni. (2015), whose see a world of possibilities at the intersection of culture, tourism, and ICT. They discuss the profound implications this convergence has on the lifecycle of cultural products – from their creation and evaluation to their management and promotion. The overarching goal is to harness digital technologies not just as tools but as catalysts that amplify the essence of cultural content, ensuring its preservation and fostering a deeper understanding. Panagiotopoulou et al. (2019) envision a future where these technological advancements bolster the allure of cultural destinations, making them more marketable than ever.

The European Commission (2022) joins this discourse, emphasizing the need to harness the potential of heritage digitization fully. This sentiment is not just about preserving the past but enhancing the cultural tourism experience of the present and a perspective shared (Buhalis & Amaranggana, 2014; Neuhofer et al., 2015). A good example of this is The Love Bank in Banská Štiavnica (for more information, see Vitálišová et al., 2022). Situated in the heart of a UNESCO World Heritage site, the museum plays a crucial role in preserving historical heritage. Its main mission is to safeguard the poem 'Marina' for future generations, thereby maintaining unique aspects of Slovak literature and history in a creative, modern format. The museum's strategy effectively combines historical heritage with innovative presentation methods. This approach proved particularly relevant during the Covid-19 pandemic, leading to new opportunities like virtual love boxes, demonstrating the dynamic potential of blending heritage with modern technology.

The post-pandemic world has further accentuated the indispensability of ICT in the realm of cultural tourism. As highlighted by Garau (2015) and Marzo-Navarro et al. (2017), the aftermath of the pandemic saw a surge in the reliance on digital platforms to keep the spirit of cultural exploration alive. Travel apps, as noted by Dickinson et al. (2014) and Xiang et al. (2015), have emerged as invaluable companions for tourists, enriching their experiences manifold. Economou (2015) further to this narrative emphasizes that these digital tools not only enhance the travel experience but also play a crucial role in deepening tourists' appreciation and understanding of local cultures and identities.

Another example of integrating technology with cultural heritage is provided by a project in the Posavje region (Slovenia), as detailed by Straus et al. (2019). This initiative aims to create an immersive digital experience for tourists, blending modern technology with the rich history and cultural significance of the region's castles. By engaging visitors in a captivating manner, the project sought not only to enhance the tourism experience but also to encourage more extensive exploration of the area's unique historical landmarks. This approach represents a novel way of promoting tourism and preserving cultural heritage using digital technology.

The European Commission emphasizes the importance of preserving and digitizing cultural heritage for the current digital age. Their Recommendation 2021/1970 promotes frameworks to bolster the cultural heritage sector, ensuring its resilience and transformation. This initiative aims to improve digitization quality, reuse, and digital preservation across the EU, benefiting sectors like tourism and research. Technologies such as Data, AI, 3D, and XR are rejuvenating cultural heritage sites. Virtual museums, for instance, offer immersive experiences, allowing visitors to view artworks in their original context. The Directorate General for Communications Networks, Content and Technology of the European Commission has been proactive in supplementing the cultural policies of Member States, focusing on digitalization, online access, and digital preservation. Social media platforms, including forums, blogs, and Instagram, are becoming key ones in promoting tourism destinations (Leung et al., 2013; Sotiriadis, 2017). Influencers on these platforms share experiences, influencing potential tourists (Moro & Rita, 2018). However, this can lead to contradictions in sustainability, as better promotion could result in over-tourism, which is not sustainable. Therefore, it is crucial to involve local communities in shaping the future of tourist destinations. Cultural tourism is undergoing a transformation, with attractions like museums adopting augmented and virtual reality to enhance visitor experiences (Richards, 2019). Digitalization is reshaping sustainable cultural tourism, making it more accessible, engaging, and informative. To conclude previous three sections, sustainable cultural tourism is characterized by its emphasis on active community involvement, the co-creation of authentic experiences, and the seamless integration of technology, as depicted in Figure 6.3. This form of tourism

*Figure 6.3* Essentials of sustainable cultural tourism.

thrives when it melds participatory methods with both age-old community engagement practices and modern technological advancements like digitalization. As the digital era progresses, the cultural tourism industry must continually innovate to offer enriched and immersive experiences for its audience.

Active community involvement emphasizes the crucial role of local communities in shaping and influencing the direction and offerings of cultural tourism, ensuring that it remains authentic and representative of the local culture. Co-creation of authentic experiences in sustainable cultural tourism encourages collaboration between tourists and locals to create genuine and memorable cultural experiences. Through integration of technology and leveraging modern technologies, sustainable cultural tourism can enhance the visitor experience, making it more interactive, informative, and accessible. Emphasis on participatory methods underscores the importance of including various stakeholders, from local communities to tourists, in the decision-making processes related to cultural tourism. Blend of traditional community engagement and modern innovations means that while it's essential to preserve and respect traditional cultural practices, sustainable cultural tourism also embraces contemporary innovations to enrich the overall experience. Continuous adaptation and innovation in the digital age are recognizing the rapid advancements in technology, and sustainable cultural tourism is always evolving, ensuring that it remains relevant and appealing to modern audiences.

## 6.5 Proposal of participatory framework for sustainable cultural tourism inspired by INCULTUM project and its pilot actions

The INCULTUM project, with its ten pioneering pilot actions, offers a groundbreaking approach to cultural tourism. The INCULTUM project, along with its pilot actions, is tailored to address both the challenges and opportunities inherent in cultural tourism. Its primary objective is to promote sustainable social, cultural, and economic growth within the territories. By tapping into the untapped potential of marginal and peripheral areas and placing their management in the hands of local communities and stakeholders, the project adopts innovative participatory methods. These approaches empower residents, transforming them into key players who can mitigate adverse effects. By learning from and enhancing best practices, these methods can be replicated and integrated into broader strategies and policies. The proposed participatory framework for sustainable cultural tourism is influenced by the insights of Panagiotopoulou et al. (2017, 2019), Borsekova et al. (2017, 2022) and the theoretical review outlined in this chapter. Drawing inspiration also from INCULTUM innovative strategies, we propose a participatory framework (see Figure 6.4) aimed at fostering sustainable cultural tourism, ensuring that both local communities and visitors collaboratively engage in preserving and celebrating cultural heritage for generations to come. Some inspiration for the proposal of this framework is drawn from INCULTUM pilot cases in Ireland and Slovakia. In Ireland, the Historic Graves project stands out as a unique, community-focused grassroots heritage initiative. It empowers local community groups by providing them with training in cost-effective, high-tech field surveys of historic graveyards and in recording their own oral histories. These groups collaborate to create a comprehensive online record of historic graves in their areas, cumulatively forming a valuable national resource. The project establishes a systematic and standardized approach to surveying historic graveyards. In Slovakia, the pilot action focuses on the underdeveloped mining heritage tourism potential in the Banská Bystrica region. It aims to create an interactive, participatory digital platform named *Mining Treasures*. This initiative is designed to engage and involve the community in preserving and promoting the region's rich mining heritage, leveraging digital tools to enhance accessibility and interest (for more information, see, Borsekova et al., 2023).

In the initial stage of our proposed participatory framework for sustainable cultural tourism, a thorough analysis of the external environment is paramount. This begins with an evaluation of how cultural tourism aligns with the global Sustainable Development Goals, emphasizing its role in holistic development. Concurrently, it's essential to understand the resilience of cultural tourism during economic downturns, particularly its adaptability and potential as a catalyst for recovery in the face of economic recessions. The European Policy Agenda's stance on the integration of culture, tourism, and digitalization is another pivotal aspect. By assessing how these elements

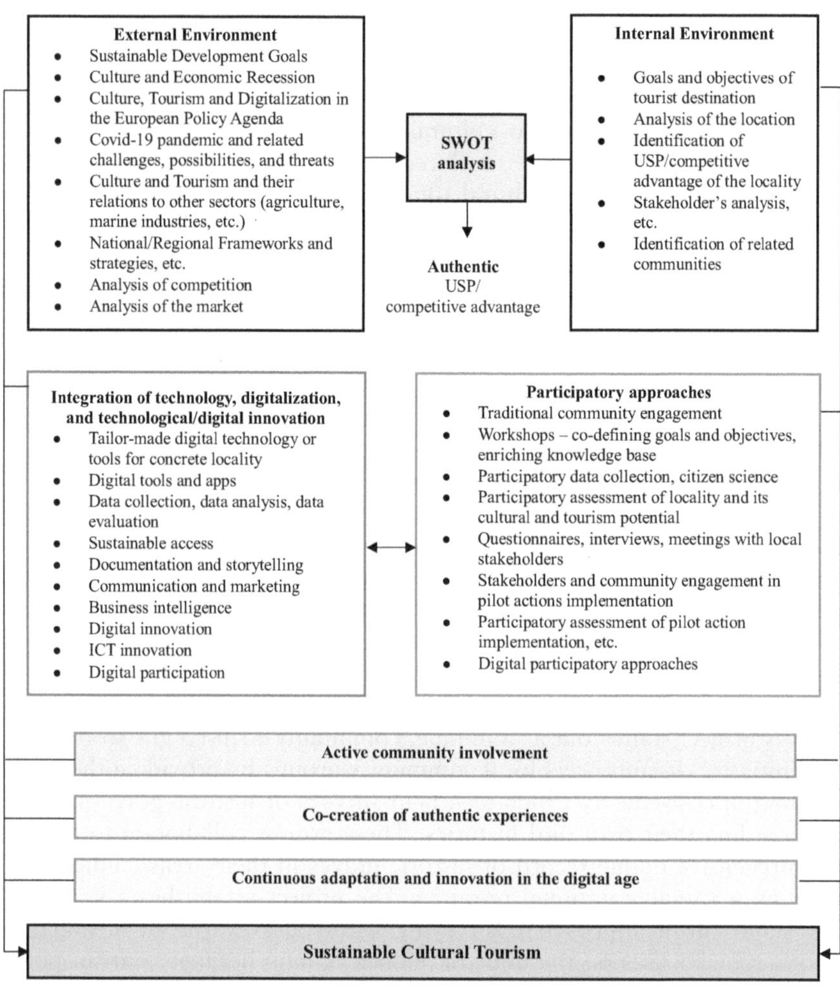

*Figure 6.4* Proposal of participatory framework for sustainable cultural tourism.

*Source:* inspired by Panagiotopoulou et al. (2017, 2019), Borsekova et al. (2015, 2017, 2022) and INCULTUM pilot actions

are prioritized and strategized by European policymakers, we can gauge the direction and emphasis of regional initiatives. The Covid-19 pandemic, with its profound impact on the global landscape, necessitates an exploration of the opportunities and threats it presents to cultural tourism. This exploration provides insights into the sector's adaptability and the innovative strategies required to navigate such unprecedented times. Furthermore, the intricate interplay between cultural tourism and other sectors, such as agriculture and marine industries, cannot be overlooked. By investigating these symbiotic relationships, we can discern how cultural tourism can benefit from and contribute to these sectors. This holistic approach is complemented by a review of existing policy frameworks and strategies at both national and regional levels, ensuring that our framework is in harmony with prevailing guidelines and objectives. Lastly, to ensure the framework's viability and relevance, it's crucial to identify the key competitors in the cultural tourism landscape. By understanding their strengths and weaknesses, we can carve out a unique niche for our framework. This is further enriched by gauging the current market dynamics, which involves understanding the preferences of the target audience and anticipating future trends in cultural tourism.

The subsequent stage of our proposed Participatory Framework for sustainable cultural tourism delves into an in-depth analysis of the internal environment. This phase begins by pinpointing the goals and objectives of the tourist destination, ensuring a clear vision and direction for future endeavours. An intricate examination of the location follows focusing on its geographical, cultural, and historical attributes that can influence tourism dynamics. Central to this stage is an identification of the Unique Selling Proposition (USPs) or the competitive advantage of the locality, which distinguishes it from other destinations and offers a compelling reason for tourists to visit. Equally vital is a stakeholder analysis, which maps out key players, their interests, and their influence on the tourism ecosystem. This comprehensive approach is rounded off by identifying related communities, ensuring that the framework is inclusive and takes into account the diverse groups that contribute to and benefit from cultural tourism. The culmination of both internal and external environment analyses results in a SWOT analysis. SWOT, an acronym for Strengths, Weaknesses, Opportunities, and Threats, provides a structured framework to evaluate tourist destination position by identifying its internal capabilities and external market dynamics. Following the comprehensive SWOT analysis, the insights gleaned from the previous steps facilitate the identification of an authentic USP or competitive advantage.

Competitive advantage, rooted in Porter's theory, is described as a superior capability to compete, central to economic and business activities in competitive markets (Porter, 1990). Two primary approaches to competitive advantage are a market-oriented approach, emphasizing external factors like market conditions (Kotler, 2002; Porter, 1990; Vaňová, 2006), and a resource-based approach, which prioritizes internal resources (Barney, 1995; Hall, 1993; Pfeffer, 1994; Powell, 1992; Ulrich & Lake, 1991). While the former leverages external opportunities, the latter underscores the significance of internal assets. This dichotomy raises debates about the true sources of competitive advantage. In tourism, 'competitive advantage' is frequently synonymous with the term 'unique selling proposition' (USP) (see, e.g., King, 2010). The essence of a USP lies in its authenticity, resonating with visitors' travel objectives and motivations. Often, authentic competitive advantage is rooted in culture and unique cultural assets that are distinct to each destination (see, e.g., Borsekova et al., 2015, 2017). This USP, rooted in genuine attributes and values of the destination, emerges as a pivotal component, becoming an integral and defining element of the entire framework. It not only differentiates the destination but also ensures its offerings resonate with authenticity and depth, enhancing its appeal to discerning tourists.

The ensuing component of our proposed framework underscores the integration of technology, digitalization, and technological/digital innovation. In the realm of their integrating, several options emerge. One can begin with the development of tailor-made digital technologies or tools, each meticulously designed to resonate with specific localities. This is complemented by the utilization of digital tools and apps, all aimed at elevating the tourist experience.

To ensure decisions are data-driven, robust systems for data collection, analysis, and evaluation are paramount. Equally vital is the sustainable access to these digital resources, ensuring longevity and reliability. The cultural essence of a destination can be vividly portrayed through innovative documentation and storytelling techniques, making the narrative come alive for visitors. In the age of digital ubiquity, communication and marketing strategies must leverage digital platforms, optimizing reach and engagement. Business intelligence tools stand as invaluable assets, offering insights into market trends and consumer behaviours, thus enabling proactive and informed strategies. To remain at the forefront, it's essential to champion both digital and ICT innovations. And, to ensure a holistic approach, promoting digital participation becomes crucial, ensuring that every stakeholder, from local communities to visitors, is actively engaged in the journey of digital transformation.

The important component of our proposed framework emphasizes the exploitation of suitable participatory approaches. This involves different approaches, for example traditional community engagement methods that foster a sense of belonging and ownership among locals. Workshops can play a crucial role, serving as platforms for co-defining goals and objectives and enriching the collective knowledge base. Participatory data collection methods, including citizen science initiatives, offer a grassroots perspective, ensuring that the data is both comprehensive and relevant. A hands-on assessment of the locality, focusing on its cultural and tourism potential, is achieved through participatory evaluations. Tools such as questionnaires, interviews, and meetings with local stakeholders provide in-depth insights and feedback. Furthermore, stakeholder and community involvement are pivotal during the implementation of pilot actions, ensuring that initiatives are grounded in local realities. The assessment of these pilot actions is also participatory, ensuring feedback loops for continuous improvement. To complement these traditional methods, digital participatory approaches are integrated, leveraging technology to enhance engagement and reach. Of course, not all of these approaches and methods need to be employed simultaneously.

It's crucial to select approaches that best align with the purpose and characteristics of the locality, ensuring they effectively support subsequent stages. As we transition to the next phase, active community involvement becomes paramount, fostering a sense of ownership and collaboration among locals. This involvement sets the stage for the co-creation of authentic experiences, where the authentic USP plays a pivotal role in crafting memorable and genuine tourist interactions. As the digital landscape continually evolves, continuous adaptation and innovation in the digital age are essential, ensuring that the framework remains relevant and forward-looking. It's vital to note that the authentic USP remains a cornerstone throughout these stages. Collectively, these steps, when executed with precision and commitment, can significantly bolster Sustainable Cultural Tourism.

## 6.6    Conclusion

This chapter set out with the primary objective of offering a thorough examination of participatory models in the realms of culture, cultural tourism, and sustainable cultural tourism. We outlined the key characteristics of cultural participation, participation in cultural tourism, and sustainable cultural tourism, emphasizing the importance of active involvement, co-creation, and the integration of technology. These elements are crucial in shaping cultural narratives, tourism products, and sustainable practices.

Active community involvement is highlighted as essential for ensuring cultural tourism remains authentic and representative of local culture. The co-creation of authentic experiences in sustainable cultural tourism is encouraged, fostering collaboration between tourists and locals to create memorable cultural experiences. The integration of technology plays a significant role in enhancing the visitor experience, making it more interactive, informative, and accessible. The emphasis on participatory methods is beneficial due to involving various stakeholders, from local communities to tourists, in decision-making processes related to cultural tourism. The blend of traditional community engagement and modern innovations suggests that while preserving and respecting traditional cultural practices is vital, embracing contemporary innovations is equally important to enrich the overall experience. Recognizing the rapid advancements in technology, continuous adaptation and innovation are necessary to ensure sustainable cultural tourism remains relevant and appealing to modern audiences.

Our in-depth exploration, enriched by the practical insights from the INCULTUM project, has culminated in the introduction of an innovative participatory framework for sustainable cultural tourism. Key characteristics of this approach include active involvement, co-creation, and the use of technology in shaping cultural narratives and tourism products.

The framework begins with an analysis of the external environment as understanding the strengths and weaknesses of key competitors and current market dynamics is very important in this stage. Internally, the framework focuses on identifying the goals, objectives, and competitive advantage or USPs of tourist destinations. The integration of technology and digitalization is a critical component of the framework. This includes the development of tailor-made digital tools, utilization of digital apps, and innovative documentation and storytelling techniques.

Participatory approaches are crucial, combining traditional community engagement methods with digital participatory approaches. Continuous adaptation and innovation in the digital age ensure the framework's relevance and effectiveness. The authentic competitive advantage or USP remains a cornerstone throughout these stages, guiding the development of memorable and genuine tourist interactions. To conclude, this framework for sustainable cultural tourism, drawing inspiration from successful pilot cases and theoretical insights, offers a comprehensive approach to fostering sustainable growth in cultural tourism. It emphasizes the importance of community participation, technological integration, and continuous innovation, ensuring a balanced and forward-looking approach to cultural tourism development.

## Bibliography


Alfarè, L., & Nared, J. (2014). Leading a participatory process. In J. Nared & N. Razpotnik Viskovíc (Eds.), *Managing cultural heritage sites in Southeastern Europe* (pp. 32–37). Ljubljana: Založba ZRC.

Arnstein, S. R. (1969). A ladder of citizen participation. *JAIP*, 35(4), 216–224.

Ateca-Amestoy, V. (2008). Determining heterogeneous behavior for theater attendance. *Journal of Cultural Economics*, 32(2), 127–151.

Ateca-Amestoy, V., & Prieto-Rodriguez, J. (2013). Forecasting accuracy of behavioural models for participation in the arts. *European Journal of Operational Research*, 229(1), 124–131.

Barney, J. B. (1995). Looking inside for competitive advantage. *Academy of Management Perspectives*, 9(4), 49–61.

Bitušíková, A. (2021) Cultural heritage and tourism in 21st century. Origin, trends and perspectives. *Signis. Banská Bystrica.*

Bole, D., Pipan, P., & Komac, B. (2013). Cultural values and sustainable rural development: A brief introduction. *Acta geographica Slovenica*, 53(2), 367–370.

Borsekova, K., Bitušíková, A., & Vitálišová, K. (2022). Report on participatory models. *Deliverable 4.1 of the INCULTUM Project.* Available at https://incultum.eu/wp-content/uploads/2023/01/INCULTUM-D4.1-new.pdf

Borseková, K., Vaňová, A.& Petríková, K. (2015). From tourism space to a unique tourism place through a conceptual approach to building a competitive advantage. In *Marketing Places and Spaces (Advances in Culture, Tourism and Hospitality Research, Vol. 10*, pp. 155–172). Leeds: Emerald Group Publishing Limited. https://doi.org/10.1108/S1871-317320150000010012

Borsekova, K., Vaňová, A., & Vitálišová, K. (2017). Smart specialization for smart spatial development: Innovative strategies for building competitive advantages in tourism in Slovakia. *Socio-economic Planning Sciences*, 58, 39–50.

Borsekova, K., Vitálišová, K., & Bitušíková, A. (2023). *Participatory governance and models in culture and cultural tourism.* Available at: https://incultum.eu/wp-content/uploads/2023/12/Book_Participatory_models_in_culture_s-obalkou.pdf

Bortolotto, C. (2007). From objects to processes: UNESCO'S'Intangible Cultural Heritage'. *Journal of Museum Ethnography*, (19), 21–33.

Botes, L., & Van Rensburg, D. (2000). Community participation in development: Nine plagues and twelve commandments. *Community Development Journal*, 35(1), 40–57.

Buhalis, D., & Amaranggana, A. (2014). Smart tourism destinations. In Z. Xiang & I. Tussyadiah (Eds.), *Information and communication technologies in tourism* (pp. 553–564). Heidelberg, Germany: Springer.

Byrd, E. T. (2007). Stakeholders in sustainable tourism development and their roles: Applying stakeholder theory to sustainable tourism development. *Tourism Review*, 62(2), 6–13.

Callot, P. (2013). Sustainable cultural tourism: Challenges and opportunities. *Journal of Tourism Challenges and Trends*, 6(2), 11–23.

Cater, E. (1994). Ecotourism in the third world: Problems and prospects for sustainability. In E. Cater & G. Lowman (Eds.), *Ecotourism: A sustainable option?* (pp. 69–86). Chichester: Wiley.

Ciolfi, L., Bannon, L. J., & Fernström, M. (2008). Including visitor contributions in cultural heritage installations: designing for participation. *Museum Management and Curatorship*, 23(4), 353–365.

Cipolla, F. F. V., Castro, L. C., Nicol, E., Kratky, A., & Cipolla-Ficarra, M. (2011). *Human-computer interaction, tourism and cultural heritage: First international workshop*, HCITOCH 2010, Brescello, Italy, September 7–8, 2010: Revised Selected Papers. Berlin, Heidelberg, DE: Springer.

Civantos, J. M. M., Rodríguez, B. R., Zakaluk, T., Ramón, A. G., & Martos-Rosillo, S. (2023). Ancestral integrated water management systems as adaptation tools for climate change: The "Acequias De Careo" and historical water management of the Mecina River in Sierra Nevada (Granada, Spain). *Conservation and Management of Archaeological Sites*, 25(1–3), 7–29. https://doi.org/10.1080/13505033.2023.2293345

Cole, S. (2006). Cultural tourism, community participation and empowerment. In E. Smith & M. Robinson (Eds.), *Cultural tourism in changing world* (pp. 89–103). Bristol, Blue Ridge Summit: Chanell View Publication.Convention. (1972). *Concerning the protection of the world cultural and natural heritage*. Paris: United Nations educational, Scientific and Cultural Organization.

Council of Europe. (2005). Framework convention on the value of cultural heritage for society. *Faro Convention*, CETS 199. [Online] Faro: Council of Europe. Available at: https://www.coe.int/en/web/cultureand-heritage/faro-convention

Dickinson, J., Ghali, K., Cherrett, T., Speed, C., Davies, N., & Norgate, S. H. (2014). Tourism and the smartphone app: Capabilities, emerging practice, and scope in the travel domain. *Current Issues in Tourism*, 17(1), 84–101.

Duncum, P. (2011). Youth on YouTube: Prosumers in a peer-to-peer participatory culture. *International Journal of Arts Education*, 9, 24–39.

Economou, M. (2015). Heritage in the digital age. In W. Logan, M. N. Craith, & U. Kockel (Eds.), *A companion to heritage studies* (pp. 215–228). Chichester, West Sussex, Malden, MA: Wiley.

Eriksson, B. (2020) ETC Casebook Participatory Theatre – A Casebook in Spring 2020. European Theater https://www.europeantheatre.eu/publication/ citizen-participation-in-arts-and-culture-birgit-eriksson, accessed 7.1.2024

European Commission (2014). *Towards an integrated approach to cultural heritage for Europe*. Communication from the Commission to the European Parliament, the Council, the European Economic and Social Committee and the Committee of the Regions. COM (2014) 47 7.

European Commission. (2019). *Directorate-general for education, youth, sport and culture, sustainable cultural tourism*. Publications Office.

European Commission. (2022). *Shaping Europe's digital future*. Available at: https://digital-strategy.ec.europa.eu/en/policies/cultural-heritage (Accessed on 17.10.2023).

Falk, M., & Katz-Gerro, T. (2016). Cultural participation in Europe: Can we identify common determinants? *Journal of Cultural Economics*, 40, 127–162.

Garau, C. (2015). Perspectives on cultural and sustainable rural tourism in a smart region: The case study of Marmilla in Sardinia (Italy). *Sustainability*, 7(6), 6412–6434.

González-Herrera, M. R., Suárez-Chaparro, R. H., & Hernández-Casimiro, K. (2022). Contribution of tourism to sustainable development: Samalayuca Dunes. In S. R. Slocum, P. Whiltshier, & J. B. Read (Eds.), *Tourism transformations in protected area gateway* communities (p. 193).Wallingford: Oxfordshire, Boston.

Gunn, C. A. (1994). *Tourism planning: Basic concepts cases* (3rd ed.). Washington, DC: Taylor and Francis.

Hall, C. M. (2017). *Slow tourism: Retracing the journey*. London: Routledge.

Hall, R. (1993). A framework linking intangible resources and capabiliites to sustainable competitive advantage. *Strategic Management Journal*, 14(8), 607–618.

Haywood, K. M. (1988). Responsible and responsive tourism planning in the community. *Tourism Management*, 9(2), 105–118.

Johnson, L., & McCarthy, J. (2019). The role of ICT in preserving intangible cultural heritage. *Harvard Journal of Tourism Research*, 12(2), 45–60.

King, A. (2010). Point of difference. *Business Review Weekly*, 36(32), 28–29.

Kiss, A. (2004). Is community-based ecotourism a good use of biodiversity conservation funds? *Trends in Ecology & Evolution*, 19(5), 232–237.

Koens, K., Postma, A., & Papp, B. (2018). Is overtourism overused? Understanding the impact of tourism in a city context. *Sustainability*, 10(12), 4384.

Kotler, P. (2002). *Marketing places*. New York: Simon and Schuster.

Leung, D., Law, R., van Hoof, H., & Buhalis, D. (2013). Social media in tourism ad hospitality: A literature review. *Journal of Travel and Tourism Marketing*, 30, 3–22.

Logan, W., Craith, M. N., & Kockel, U. (Eds.) (2015). *A companion to heritage studies*. Chichester, West Sussex, Malden, MA: Wiley-Blackwell.

Lourens, M. (2007). Role of wellness tourism in health promotion. *African Journal for Physical Health Education, Recreation and Dance*, 13(3), 255–268.

Luccetti, V. G., & Font, X. (2013). Community based tourism: Critical success factors. *ICRT Occasional Paper*, 27, 1–20.

Margiono, A. (2021). Digital transformation: Setting the pace. *Journal of Business Strategy* 42(5), 315–322.

Marzo-Navarro, M., Pedraja-Iglesias, M., & Vinzón, L. (2017). Key variables for developing integrated rural tourism. *Tourism Geographies*, 19(4), 575–594.

Matteucci, X., & Von Zumbusch, J. (2020). *Theoretical framework for cultural tourism in urban and regional destination*. Deliverable D2.1 of the Horizon 2020 Project SmartCulTour (GA Number 870708), Published on the Project Web Site on July, 2020. Available at: https://Www.Smartcultour.Eu/Deliverables/

McKercher, B., & du Cros, H. (2002). *Cultural tourism: The partnership between tourism and cultural heritage management*. New York: Routledge.

Mergel, I., Edelmna, N., & Haug, N. (2019). Defining digital transformation: Results from expert interviews. *Government Information Quarterly*, 36(4), 1–16.

Moro, S., & Rita, P. (2018). Brand strategies in social media in hospitality and tourism. *International Journal of Contemporary Hospitality Management*, 30(1), 343–364.

Mossberg, L. (2008). Extraordinary experiences through storytelling. *Scandinavian Journal of Hospitality and Tourism*, 8(3), 195–210.

Murphy, P. (1985). Tourism: a community approach. Methuen: New York and London. *Tourism Review*, 6(2), 9–10.

Nared, J. (2014). Participatory planning. In J. Nared & N. Razpotnik Visković (Eds.), *Managing cultural heritage sites in southeastern Europe* (pp. 28–32). Ljubljana: Založba ZRC.

Nared, J., & Bole, D. (2020). Participatory research on heritage- and culture-based development: A perspective from South-East Europe. In: J. Nared, & D. Bole (Eds.), *Participatory research and planning in practice*. The Urban Book Series. Cham: Springer.

Nared, J., Erhartič, B., & Visković, N. R. (2013). Including development topics in a cultural heritage management plan: Mercury heritage in Idrija. *Acta Geographica Slovenica*, 53(2), 394–402.

National Trust for Historic Preservation. (2001). *Share your heritage: Cultural and heritage tourism: The same or different?* Washington, DC.

Neuhofer, B., Buhalis, D., & Ladkin, A. (2015). Smart technologies for personalized experiences: A case study in the hospitality domain. *Electronic Markets*, 25(1), 243–254.

OECD. (2021) *Culture, creative sectors and local development*. Policy webinar series, Available at: https://www.oecd.org/cfe/leed/NEXT-TO-FINAL-AGENDA-OECD-EC-Webinar-1-3-%20December-cultural-participation.pdf (Accessed on 5.10.2023).

Ozcevik, O., Beygo, C., & Akcakaya, I. (2010). Building capacity through collaborative local action: Case of Matra REGIMA within Zeytinburnu regeneration scheme. *Journal of Urban Planning and Development*, 136(2), 169–175.

Panagiotopoulou, M., Somarakis, G., Stratigea, A., & Katsoni, V. (2017). In search of participatory sustainable cultural paths at the local level—the case of kissamos province-crete. In V. Katsoni, A. Upadhya, & A. Stratigea (Eds.), *Tourism, culture and heritage in a smart economy* (pp. 339–363). Springer Proceedings in Business and Economics. Cham: Springer. .

Panagiotopoulou, M., Somarakis, G., & Stratigea, A. (2019). Participatory planning in support of resilient natural/cultural resource management. In A. Stratigea, & D. Kavroudakis (Eds,).*Mediterranean cities and Island communities* (pp. 181–211). Cham: Springer.Peng, P., & Dewa Putu, O. P. (2011). Community participation for sustainable tourism in heritage site: A case of Angkor, Siem Reap Province, Cambodia. *Mudra*, 26(3), 1–1.

Pfeffer, J. (1994). Competitive advantage through people. *California Management Review*, 36(2), 9–28.Porritt, J. (1998) Foreword. In D. Warburton (Ed.), *Community and sustainable development* (pp. xi–xiv).London: Earthscan.

Porter, M. E. (1990). The competitive advantage of nations. *Harvard Business Review*, 68(2), 73–93.

Potts, J., Hartley, J., Banks, J., Burgess, J., Cobcroft, R., Cunningham, S., & Montgomery, L. (2008). Consumer co-creation and situated creativity. *Industry and Innovation*, 15(5), 459–474.

Powell, T. C. (1992). Organizational alignment as competitive advantage. *Strategic Management Journal*, 13(2), 119–134.

Richards, G. (2018). Cultural tourism: A review of recent research and trends. *Journal of Hospitality and Tourism Management*, 36, 12–21.

Richards, G. (2019). Culture and tourism: Natural partners or reluctant bedfellows? A perspective paper. *Tourism Review*, 75(1), 232–234.

Robson, J., & Robson, I. (1996). From shareholders to stakeholders: Critical issues for tourism marketers. *Tourism Management*, 17(7), 533–540.

Rodriguez, P., & Moretti, A. (2020). Smart cities and cultural tourism: A future perspective. *Harvard Review of Urban Planning*, 5(1), 34–49.

Sacco, P., Ferilli, G., & Blessi, G. T. (2013). Understanding culture-led development: A critique of alternative theoretical explanations. *Urban Studies*, 51(2), 806–821.

Sacco, P. L., Ferilli, G., & Tavano Blessi, G. (2018). From culture 1.0 to culture 3.0: Three socio-technical regimes of social and economic value creation through culture, and their impact on European Cohesion Policies. *Sustainability*, 10(11), 3923.

Salazar, N. B. (2012). Tourism imaginaries: A conceptual approach. *Annals of Tourism Research*, 39(2), 863–882.

Seifert, A., & Rössel, J. (2022). Digital participation. In: D. Gu, & M.E. Dupre (Eds.), *Encyclopedia of gerontology and population aging* (pp. 1446–1450). Cham: Springer International Publishing.

Simpson, M. C. (2008). Community benefit tourism initiatives—A conceptual oxymoron? *Tourism Management*, 29(1), 1–18.

Slocum, S. L., Aidoo, A., & McMahon, K. (2020) *The business of sustainable tourism development and management*. Oxon: Routledge.

Smith, A., & Richards, G. (2013). *The Routledge handbook of cultural tourism*. London: Routledge.

Smith, L. (2006). *Uses of heritage*. London: Routledge.

Smith, M. K. (2009). *Issues in cultural tourism studies*. London: Routledge.

Smith, M. K. (2016). *Issues in cultural tourism studies*. London: Routledge.

Sonkoly, G., & Vahtikari, T. (2018). *Innovation in cultural heritage research*. For an Integrated European Research Policy. Brussels: European Commission.

Sotiriadis, M.D. (2017). Sharing tourism experiences in social media: A literature review and a set of suggested business strategies. *International Journal of Contemporary Hospitality Management*, 29(1), 179– 225.

Stratigea, A., & Katsoni, V. (2015). A strategic policy scenario analysis framework for the sustainable tourist development of peripheral small island areas—the case of Lefkada-Greece island. *European Journal of Futures Research* (EJFR), 3(5), 1–17.

Straus, M., Starc-Peceny, U., & Celgar, K. (2019) *Digital innovation of cultural heritage*. Handbook for Tourist Destinations and Cultural Heritage Institutions. Ministry of Economic Development and Technology of Slovenia.

Thompson, M., Lee, D., & Green, R. (2018). Community engagement in tourism: A participatory approach. *Journal of Sustainable Tourism*, 26(7), 1101–1118.

Timothy, D. (2011) *Cultural heritage and tourism: An introduction*. Bristol: Channel View Publications.

Timothy, D. J., & Nyaupane, G. P. (2009). *Cultural heritage and tourism in the developing world: A regional perspective*. New York: Routledge.

Tosun, C., & Timothy, D. (2003). Arguments for community participation in the tourism development process. *The Journal of Tourism Studies*, 14(2), 2–15.

Tubadji, A. (2012) Culture-based development: Empirical evidence for Germany. *International Journal of Social Economics*, 39(9), 690–703.

Turner, R., & Jenkins, O. (2021). Participatory approaches to tourism development in Komodo National Park. *Environmental Stewardship and Cultural Engagement*, 7(3), 210–225.

Ulrich, D., & Lake, D. (1991). Organizational capability: Creating competitive advantage. *Academy of Management Perspectives*, 5(1), 77–92.

UNESCO. (2009). Measuring cultural participation. Framework for Cultural Statistics Handbook No. 2. Available at: https://uis.unesco.org/sites/default/files/documents/measuring-cultural-participation-2009-unesco-framework-for-cultural-statistics-handbook-2-2012-en.pdf. (Accessed on 5.10.2023).

Vaňová, A. (2006). *Strategic marketing planning of territorial development*. Banská Bystrica: Faculty of Economics UMB.

Vernon, N., Essex, S., Pinder, D., & Curry, K. (2005). Collaborative policymaking: Local sustainable projects. *Annals of Tourism Research*, 32(2), 325–345.

Vial, G. (2019). Understanding digital transformation: A review and a research agenda. *The Journal of Strategic Information Systems*, 28(2), 118–144.

Vijayanand, S. (2013). Stakeholders and public private partnerships role in tourism management. *International Journal of Scientific & Engineering Research*, 4(2), 1–11.

Vitálišová, K., Borseková, K., & Vaňová, A. (2022). Digital transformation in the preservation of cultural heritage in cities: The example of love bank. In F. Calabrò, L. Della Spina, & M. J. Piñeira Mantiñán (Eds.), *New Metropolitan perspectives. NMP 2022.* Lecture Notes in Networks and Systems (vol. 482, pp. 2617–2627). Cham: Springer. https://doi.org/10.1007/978-3-031-06825-6_250

Wang, N. (1999). Rethinking authenticity in tourism experience. *Annals of Tourism Research*, 26(2), 349–370.

Wang, Y. C., & Fesenmaier, D. R. (2007). Collaborative destination marketing: A case study of Elkhart Country, Indiana. *Tourism Management*, 28, 863–875.

Wild, C. (1994). Issues in ecotourism. In: C.P. Cooper, & A. Lockwood (Eds.), *Progress in tourism, recreation, and hospitality management* (vol. 6, pp. 12–21). Chichester: Wiley.

WTO, UNEP. (2005). *Making tourism more sustainable: A guide for policy makers*. Paris: United Nations Environmental Programme.

Xiang, Z., Wang, D., O'Leary, J. T., & Fesenmaier, D. R. (2015). Adapting to the Internet: Trends in travelers' use of the web for trip planning. *Journal of Travel Research*, 54(4), 511–527.

# 7 Tourism and reception of visitors as a lever for inclusiveness and resiliency of heritage communities

## Two case studies in Hungarian peripheries

*Eszter György, Gábor Oláh
and Gábor Sonkoly*

Due to the recent and indeed permanent critique of the scientific utility of resilience, it seems to be losing its scientific relevance and can rather be used to reflect on policy needs. It seems to be moving back and forth between the political and scientific spheres. The question arises: embracing resilience in policy, definitely loses its scientific relevance. In what follows, we attempt to examine how it can still be relevant in research, specifically in the field of cultural heritage in two Hungarian cases, which differ in their scale as destinations for cultural tourism, but they share two significant characteristics. First, they are situated in the inner peripheries of Hungary: one is within Budapest, and the other is in the vicinity of a small village of c. 750 inhabitants in Southern Transdanubia (see Figure 7.1). Second, they both demonstrate how a minority, which can be ethnic or religious, could find the means of survival and self-representation in current Hungary. We intended to understand the evolution of the two cases through a critical model of resilient communities, which proved to be not only effective in the depiction of the importance of third-regime cultural heritage in the identity-building of minorities but also in making these attempts comparable for further research. In this way, these cases are also relevant to understanding and highlighting how local communities may participate and cooperate in engaging in sustainable cultural tourism strategies.

## 7.1 Cultural heritage as the embedded potential of community resilience: from political to scientific relevance

As Humbert and Joseph (2019) aptly remarked 'the policy discourse often gives the impression that resilience has become a buzzword; making it easy to dismiss as being overused and lacking substance; as lacking conceptual weight or depth' (p. 217). It is noticeable, on the one hand, that it is

DOI: 10.4324/9781003422952-7

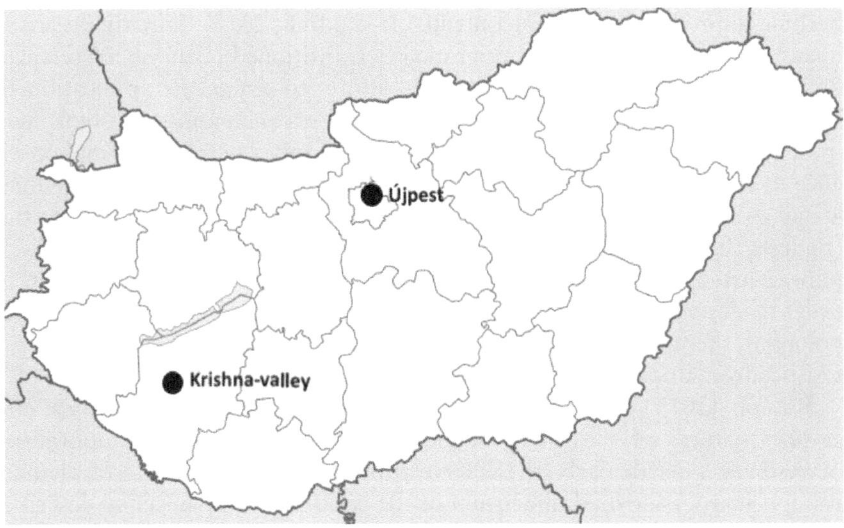



*Figure 7.1* The location of the two case studies in Hungary.

increasingly used in policy documents; on the other hand, most social science literature refers to it as a boundary concept that is holistic, interdisciplinary, situational, relational, and selective; at the same time, a significant number of this literature denote that it is poorly defined, shallow and elusive, and seemingly suitable for describing contradictory phenomena (Joseph, 2013; Walker & Cooper, 2011; Wilson, 2017).

In most cases, resilience is associated with crisis. Walker and Cooper (2011) attribute the growing popularity of resilience to the terror attack of 9/11 or the financial crisis of 2008 which introduced 'new methods of futurology, contingency planning, and crisis response onto the policy reform agenda' (p. 152). Concisely, the theory – developed in a scientific milieu – has discovered its way into political economy and security policy.

When the policy discourse on risk and security found the concept of resilience, it saw *a priori* as a possible form of governance (Joseph, 2013), especially at the local level (Bollig, 2014). It allowed formulating questions on the transition of a system: when and how does resilience come into play in change? These political efforts to define modalities of aftershock system transition fundamentally emanated from the complexity sciences of the 1970s (Walker & Cooper, 2011). Without going into more detail on the scientific conceptual history of resilience – which many have already done before us – we would like to reflect on certain aspects of complex system theories that allow us to see how resilience found itself in the transition between scientific and policy discourse.

The concept emerged from critiques of predictive and deterministic models. According to its fundamental critique of modernity, the future has become unpredictable and uncontrollable (Endreß, 2019). Initially theorised in systems ecology, this concept gradually transitioned to the social sciences during the 1990s, driven by deliberate efforts to extend its application to the modeling of social systems. Integrating society, economy, and biosphere in one socio-ecological model held the opportunity to create a complex science of resilience (Walker & Cooper, 2011). The concept increasingly infiltrated the study of all kinds of social, mental, or cultural units and entities (Endreß, 2019). Resilience started to influence the disciplines and the disciplines started to think about the meaning and applicability of the concept. In order to develop and maintain its scientific use, applicability, and analytical potential, there is a constant demand to clearly distinguish one of its characteristics from another (Meyen & Schier, 2019).

Rampp (2019) drew attention to the distinction between *ex-ante* and *ex-post* natures of the theory while examining the varying temporalities of resilience, particularly the interrelationship of continuity and change. Rampp stresses the dual dimension of the concept of resilience: he describes the forward-looking, learning, and adaptive actions of communities as an *ex-ante* process, while the efforts to manage the crisis and recovery as an *ex-post* process. While in the former it appears as management of change/ transition, in the latter as coping with disturbance. By reviewing the literature, Böschen, Binder, and Rathgeber (2019) identify and categorise the main definitions of resilience. On the one hand, they distinguish between interpretations as structure (statics) and process (dynamics); on the other hand, they contrast contextual and self-resilience. These pairs of opposites give four categories in a matrix that they consider suitable for determining the areas of resilience scientific applicability.

Brand and Jax (2007) distinguish the descriptive, hybrid, and normative nature of the resilience application in scientific literature. They argue that it was initially introduced in system ecology, sociology, and ecological economics as a description of system property or capacity. In its hybrid understanding, the concept presupposes a way of thinking that lacks any well-defined model, but rather a set of ideas about the interpretation of a system and the selection of desirable values. The latter is already in the field of normativity where resilience appears as a guideline for a continuous management of change and maintaining basic system functions. According to Endreß (2019), resilience by being a normative reference point and essentially conceptualised as a positive condition can undermine its analytical utility. In contrast, according to Bollig (2014), resilience's hybrid and normative nature fuels cross-disciplinary exchanges and facilitates communication between academia, politicians, and practitioners. However, beyond this transitional ability, its use on an empirical level is still questionable.

With the efforts to create guidelines and standardise the principles and modalities of resilient governance, Walker and Cooper (2011) argue that the basic theoretical stance of resilience science transformed into a fully operative management methodology. They point out that as the credibility of resilience as a theoretical model decreased, its depoliticising and neutralising properties/functions became more predominant in the context of neoliberalism or security policy.

Joseph (2013) states that the enthusiasm surrounding resilience is due to the fact that it easily fits and can be inserted into neoliberal policies. In his argument, by its seemingly new form of governance, resilience promotes self-organisation and self-responsibility and envisages non-intervention by the state in order to prepare individuals and local social groups to an age of uncertainty. Through its responsibilisation effort, individuals and communities are supposed to define their way of governance more autonomously to be able to show adaptability (Humbert & Joseph, 2019; Joseph, 2013; Mckeown & Glenn, 2018).

In some respects, the model of the adaptive cycle and its multi-scale version, the panarchy, was created in order to integrate the social sciences into a general systems theory (Bollig, 2014; Walker & Cooper, 2011). Understanding change has become a primary assignment for social sciences, which is in the model 'conceptualised as inevitable, discontinuous, but at the same time phased and patterned' (Bollig, 2014, p. 259). Rampp (2019) considers that the adaptive cycle model unlocks the classic historiographical dichotomy between continuity and change and makes it an interdependent and reciprocal relationship: 'change being a prerequisite for continuity addresses the change of (as the case may be: central) elements of an object, while preserving the nucleus of this object' (p. 59). Anthropologists, archaeologists, geographers, etc. established research programs to work with long timescales to examine social and cultural structures historically which opened up the possibility of revisiting periods with profound transformation (Bollig, 2014).

Resilience has therefore a fairly broad perspective on temporality. Hence, Obrist, Pfeiffer, and Henley (2010) highlight that resilience research should focus on proactive (*ex-ante*) attitudes, on the extent to which communities can learn from past experiences, from 'the stock of experience available' (p. 290), and use them to shape future strategies. This underlying temporality basically gives the concept a presentist atmosphere, and consequently, resilience becomes easily interpreted in the context of concepts like memory, commemoration, identity, or heritage (Hartog, 2015; Sonkoly, 2017).

In terms of interoperability between policy and research, heritage and resilience are analogous concepts in many ways. Due to the paradigm shift of the third regime, the concept of cultural heritage has become more complex which has been reflected in both scientific and policy discourses. The third regime of cultural heritage is characterised by conceptual novelties such as community heritage, cultural landscape, cultural rights, historical urban landscape, intangible heritage, etc. (Bendix, Eggert, & Peselmann, 2017; Smith, 2010; Sonkoly, 2017). In this transformed perspective, cultural heritage is considered to have a major impact on community resilience by representing other social and economic values and objectives in its management (Lazzeretti, 2012; György & Oláh, 2018). Resilience is therefore stimulating debate on values in terms of selecting and prioritising them (Meyen & Schier, 2019).

With the paradigm shift described as the third regime, communities are supposed to define their cultural heritage more autonomously within a bottom-up approach. On the one hand, this attitude fits with the appeal of responsibilisation, as it puts the (re)use, the preservation, and the management of heritage into the governance of the local community, which also formulates goals beyond the specific values to be preserved (Sonkoly & Vahtikari, 2018). On the other hand, institutionalising community management of heritage at a local level is increasingly also a manifestation of political will, positioning itself in relation to the mainstream canon (György & Oláh, 2018).

Undoubtedly, the locality is strongly embedded in resilient governance, as Chandler (2014) explains. The model emerged principally as a rejection of the modernist, top-down methods, which therefore vindicate a bottom-up approach for resilient governance. The object of governance – in this case, third regime cultural heritage – is conceived as complex or rather holistic. Moreover, 'there is no separation of governance from the object of governance' (p. 62) so resilient heritage communities can be understood as governing *per se* entities. Chandler introduces the concept of 'everyday democracy' in relation to resilience which refers to the level of connectedness and relational abilities of the local population. This is also emphasised by Beel et al. (2017) with the addition that the central issue to consider is the role of human agency in the process of continuous recreation of cultural heritage. If resilience is understood as human agency, at first glance two directions can be outlined at the research agenda level: (1) the historical process of re-creating and maintaining cultural heritage in the long term; (2) the modalities of setting up connectedness, the topologies of relational abilities, and the institutionalisation of heritage stewardship (György & Oláh, 2018) linked to the management of cultural heritage.

Suppose we carry forward this idea that in the paradigm of the third regime, cultural heritage is interpreted as constantly changing, and the engine of these changes is the community itself. In that case, we can grasp that resilience in this is definitely designated as internal or intrinsic. Thus, the resilient cultural heritage governance approach is unquestionably bottom-up and participative, support and integrative approaches are emerging from the community potential, not from external actors or ideas. Potentials are embedded in the system that differences in exploitation methods or strategies set it apart from others and could explain socio-spatial disparities.

While taking into consideration the critical standards of social sciences, bringing all this to the research agenda is not without its challenges as resilience goes along with a constant political/ideological burden and is linked to identity formations. A concept or model considered elusive, bordering, or even neoliberal; which attempts to make the natural and social sciences interoperable and, in some interpretations, promotes a kind of dynamised stability; is largely used as a political reference.

Our aim is not only to reflect on the concept as political, but also to unfold its scientific relevance. By approaching communities from the perception of their own resources, we can consider it as an important step to make the concept operational and analytical at the same time. This framework considers resilience as a process that reflects and, to some extent, revisits the expansion and transformation models of Böschen, Binder, and Rathgeber's (2019) matrix. In the following, attempts will be made to assess the scientific relevance of resilience based on two case studies in Hungary, one in the urban and one in the rural periphery, with ethnic and religious minority communities.

## 7.2   Roma resilience: resistance, adaptation, and participatory cultural heritage

In the following section, we are introducing the concept of resilience in Romani studies and its interconnectedness with sustainable, participatory cultural heritage-making. To illustrate different strategies for the social and cultural integration of Roma communities and the mechanisms of cultural resistance and adaptation concerning the political-cultural structures offered by the majority society, we are presenting in the second part of this section the example of a local Roma heritage centre.

Resilience is often referred to as the ability of different Gypsy/Traveller/ Romani groups to adapt cultural practices and identities to new environments. Despite the lack of cultural continuity and the very heterogeneous cultural/ethnic/social characteristics of Romani groups throughout Europe, several studies are proving the presence of strong adaptive practices and cultural resilience in the face of assimilatory pressures (Greenfields & Smith, 2018). Some authors explain the need for adaptive strategies to the lack of nation-states and acknowledge historical identity, stating that 'the Gypsies have constantly had to adjust their histories and identities in the interest of survival' (Scott, 2009, p. 235). Others stress on the resilient nature of cultural mechanisms:

> Gypsy culture is created through contact, sometimes conflict and specific exchange. Gypsy culture is one emerging from ever-present and changing culture contact rather than a former isolate allegedly undermined by contact. Theirs is a culture created from and through difference.
>
> (Okely, 2012, pp. 40–41)

In other respects, resilience comes up with regard to the various relationships and positions of Roma as minorities, facing the non-Roma majority societies. Specific ways of resilience and adaptation during state-socialist regimes were analysed, showing how cultural norms and attributes survived and evolved despite harsh assimilatory policies. Stewart (1994) in his book *Brothers in*

*Song* constructed the equation of (Gypsy) + (socialist wage work + housing) = (Hungarian worker) + (Gypsy folklore), justifying how could, paradoxically and despite the required assimilation, Vlach Gypsy groups in Hungary preserve their cultural heritage. According to Stewart, they were 'salaried workers on weekdays, Gypsy on Sundays', who, despite the fact of not being considered a nationality, 'are not ceasing to exist but reproduce themselves in the frames of the Hungarian social and economic circumstances with the same resoluteness as ever before' (p. 71).

Silverman (2014) found cultural resistance in the elusion of the Bulgarian socialist state, which consciously targeted Romani music in its ethnonationalist cultural project of 'Bulgarisation'. In her book *Romani Routes*, she revealed how in the 1980s, when the specific woodwind instrument called *zurna* was prohibited from all contexts (festivals, media, urban, and village celebrations) – because of its Turkish descent –, musicians continued to play it as wedding music. According to Silverman, Bulgarian Roma defied the state for economic reasons and not for moral imperatives; moreover, their everyday strategies altered between accommodation and resistance to the state. Roma music survived in Bulgaria and in the postsocialist context and became a significant part of the postmodern world music market. Besides the worldwide popularity of Romani music from Bulgaria and the Balkans, the openness and hybridity of Romani music are stated in a lot of other cases, stressing the adoption of non-Romani and multiple Romani styles as well, creating artistic bricolages (Okely, 2012; Silverman, 2014).

From Silverman's example, it is also palpable that the resilience of Romani communities signifies the relationship or the coping mechanisms with the (often hostile) state or majority society. As Rampp (2019) states, resilience from a sociological point of view means the capacity to change in order to maintain the same identity and it comprehends three basic capacities: coping, adaptation, and transformation. In Acton's (1974) typology of Gypsy/Traveller resistance to state control, we find four key modes of adaptation: (1) the conservative approach, minimising contact or withdrawing; (2) cultural disintegration; (3) passing, competing by disguising ethnicity; and (4) cultural adaptation or the above-mentioned bricolage. From this classification, only the fourth mode offers favourable strategies, enabling positive outcomes for the Roma community. According to Greenfields and Smith (2018, p. 78), this 'flexible adaptation represented by the recreation of traditional communities in a new context is a form of cultural resilience, which in the context above can be perceived as encompassing active resistance to externally imposed assimilatory pressures'.

Another framing describes the relationship between majority societies and Romani groups, from the point of view of resilience, as a set of symbiotic, yet permanently rival positions, such as two competing 'weather fronts'. As stated by Allen (2018), the first weather front lies in the hostile and stigmatising behaviour of the majority society, encouraging blame and causing self-determined social isolation and rejection of public services and mainstream community norms. This weather front also emerges in public discourse through hate speech memes, prohibiting acceptance and inclusion of Romani groups. The second weather front comprises the Romani people's response to the mechanisms of social rejection and hostile discourse, but it is sustained by enacted cultural resilience. According to Allen, this second weather front is less influential than the first, but it might empower Romani communities to minimise discrimination, assimilation, and respond proactively to the threat of social rejection. 'Most importantly, this second weather front enables victimised people to retain cultural independence, and it provides a growing opportunity for a social movement that enables the voices of Romani and Traveller people to emerge' (p. 10).

This latter thought leads us to the following problematics of coping mechanisms, residing in the domain of 'Roma inclusion', comprehending civil rights movements and various policies addressed by national and international authorities. Baar (2011) explains that no other group has undertaken so many different inclusion, empowerment, and development programs than Romani groups all around Europe, including social inclusion programs by the EU, Decade Action Plans, Open Society Institute, World Bank, and national action plans. These programs aimed at creating opportunities for social, political, and cultural self-articulation and created an entire enterprise, defining new European narratives of inclusion, integration, and community cohesion. However, several (Roma) authors accused these policies of being paternalistic, tokenistic, and closely tied to patron-client relationships. Moreover, as Rostas and Rövid (2015) describe, the participation of Roma actors in these programs did nothing but legitimise the existing structures, while also establishing a particular market of 'Roma activists'. Trehan and Kóczé (2009) refer to the racialised hierarchy, – maintaining post-imperial privileges – of the 'Roma industry' in the post-socialist context. According to her, initiatives of Roma inclusion after the collapse of state-socialist and communist regimes paradoxically nourished the marginalisation of Roma communities, by keeping them at the bottom rung of a racialised hierarchy. This hierarchy consists of (principally American) Western advocacy elites at the top, below them, there are Eastern European and still non-Roma elites who are followed by Roma (mostly urban, educated) elites and at the bottom, there are finally the local Roma communities (with rural and less educated representatives).

It seems that in both cultural and political domains, Roma communities are up to this day exposed to neocolonial and oppressive dynamics. As Ruethers (2013) puts it, in the post-socialist context,

> Roma are presented as a social problem. Research is divided between research on the Roma as a migration or minority problem and research on the Gypsy craze, East European Gypsy musicians, and the post-colonial mechanisms of the Western music market.
>
> (p. 685)

If we confront these arguments related to the ongoing (more or less explicit) hostility and suppression of the majority societies to the above-mentioned theories of resilience, we may state that the resistance of Roma groups lies in the mixture of continuous adaptation and the retention of a certain autonomy. On the one hand, it is discoverable in artistic creativity and cultural production, referring principally to the statements of Okely (2012) about the semi-autonomous cultural space of Roma, creating and recreating cultural autonomy and constructing authenticity with the historic ingenuity and inventive originality amid others' cultures. On the other hand, one may detect the recent development and institutionalisation of Romani scholarship under the slogan of 'nothing about us without us', supported by – among other examples – the creation of the Romani Studies Program at CEU[1] in 2017, the funding of European Roma Institute for Arts and Culture[2] in 2017, and the publishing of Critical Romani Studies journal in 2018.[3]

If we look at resilience in relation to cultural heritage, we may find a third path, between resistance and dissolution. As mentioned before, cultural heritage in the third regime is very much linked to communities and participation; therefore, it may offer novel ways of adaptation and survival. With the active involvement of communities, the maintenance and re-use of cultural heritage are established 'from below', manifesting often counter-hegemonic practices (Beel et al., 2017, p. 462). However, the resilience of the cultural heritage of minorities/Roma groups should not be understood in an isolated manner of cultural survival but in the semi-autonomous, culturally hybrid space, as described by Okely (2012). Despite the long history of discrimination and oppression of European Roma, their history and heritage can only be interpreted as part of the national and transnational narratives. As Jake Bowers wrote about The Surrey Project, a specific example of preserving Roma heritage in the UK:

> The way a society treats its minorities is a litmus test of its civilisation. – so, the refusal to accept Roma history into the mainstream is also a reflection of how willing the British society is to face up the questions of who we are and what we have done.
>
> (Bowers, 2009, p. 28)

Examples of the resilience of Roma cultural heritage and their potential linkages to cultural tourism in Hungary may be rare and detached from each other but what is common in them is precisely the role of the local community and the fine – and often hardly achieved – balance between community participation and relationship with the surrounding institutions and actors of the majority society. In the last part of this section, we present briefly a unique example of resilient cultural heritage, the Roma local history collection in Újpest, where the aforementioned weather fronts seem to consolidate with each other.

'Why is that the Gypsy kid is throwing more garbage on the street than the Hungarian one? Because he doesn't feel the city as his own'[4] said István Gábor Molnár to a (presumably racist) question of a local politician in the late 1990s during a municipality meeting in Újpest, in the 4th district of Budapest. This discriminative yet everyday experience encouraged the young leader of the Roma local government and teacher in a segregated primary school to launch exceptional research to discover the history of the Roma community in Újpest.[5] His principal aim was to prove that these people belonged to the city as much as their non-Roma neighbours and that since the foundation of Újpest, through its industrial development until its annexation to the capital, Roma people had been an integral and active part of the local society. The research was conducted in two parallel ways: local historians – who in general did not have much knowledge about Gypsies living in the neighbourhood – were asked to collect data about Roma people in the archives and the media, while István Gábor Molnár started to interview the inhabitants, first of all, the elderly Roma Musicians of the district. The traditional and oral history research was in a lot of cases contradictory and highlighted the discrepancy between an official canon (often trying to erase the presence of minorities) and the personal and communal memory of the Roma people of Újpest. Thus, an innovative methodology was established, relying on the active participation of the local Roma families and resulting in the collection of more than 3000 photos, several hundreds of archival documents, and family interviews. Finally, the investigation not only proves that the local Roma community has always been an organic and important part of urban society but also succeeded in involving families and making their life and cultural heritage important, as well as reinforcing the sense of belonging to their place of living.[6]

Today, the local Roma history collection of Újpest[7] – located in the middle of a peripheral housing estate built during the state-socialist era – stands as one of the very few 'living' Roma heritage sites in Budapest. Besides a small number of public statues and street names, commemorating Roma artists (mainly musicians), there are no other sites in the Hungarian capital where the local history and cultural heritage of the largest minority group would be permanently maintained and exhibited. Carrying out a great number of functions, the centre displays the permanent exhibition of the local Roma history collection, showing tangible heritage elements such as family photographs, and traditional craft tools of nail smithery, and offers a study hall with after-school learning activities and sporting facilities. It features the memory room of Gábor Dilinkó, Roma painter and writer, one of the few Roma heroes of the revolution of 1956, and in front of the building, a *Stolpersteine* (stumbling stone) installed in 2015 to remember József Dráfi, a member of one of the first Roma families settled in Újpest who was killed in Ravensbrück, on 20 April 1945. Moreover, the centre continuously builds a Roma library (comprising Roma literature, scientific volumes, and dictionaries in Hungarian, Romani, and other languages), also with the active participation of Roma and non-Roma donors. The building space is very often used for intimate family gatherings such as weddings, baptizers, and even public wakes (having traditionally a specific importance and length in Roma culture). Finally, the community space has an integrative function as well: it does not only serve the Roma minority but the local non-Roma population as well: for instance, here gather school directors for official ceremonies and the neighbouring houses for their general assembly.

As Nasture (2015, p. 30) states:

> There is no alternative open to Roma other than mobilising communities to seek political and civil power. The change that needs to be generated is in the attitude of the people, namely shifting from being passive to active citizens who become masters of their own destiny.

The Roma Centre in Újpest empowers the community and therefore builds local heritage and memory as a result of adapting and coping with the locality. The resilient governance of this place lies in the ingenious aggregation of cooperation, resistance, and self-representation.

### 7.3   Hungarian Society for Krishna Consciousness in Hungary: resistance and adaptation

In the last section, we are presenting a religious minority in Hungary, which could not only maintain its position as a recognised religious organisation despite its unfavourable political reception in the early 1990s, but it also founded a rural community within the inner peripheries of the country, which has become a popular tourist destination in the last 30 years. The demonstration of such resilience will be studied in the interconnectedness between this religious organisation and its rural community – Krishna valley – which is simultaneously its showcase for a wide public and the example of seclusion from contemporary consumer society.

Krishna Valley is a farm community established by the Hungarian Society for Krishna Consciousness (HSKCON) in 1993. The HSKCON is a Hungarian branch of ISKCON (International Society for Krishna Consciousness) founded in 1966 in New York by Bhaktivedanta Swami Prabhupāda (1896–1977), a Bengali scholar and missionary, who travelled to the USA, and later to another 24 countries at his elderly age, to make Gaudiya Vaishnavism a worldwide religious and cultural movement. Vaishnavism is one of the most ancient and most well-known forms of Hinduism, which was reformed by Sri Caitanya Mahaprabhu (1486–1534) – hence the denomination Gaudiya referring to Western Bengal, where he was born and active during his youth. The principal message of this reform was to make the devotional love of God, in his most intimate and original form as 'Krishna', available for everyone regardless of one's religion, ethnicity, social/economic status, caste, or gender (Tóth-Soma, 1996). Bhaktivedanta Swami propagated this reform internationally. Although he introduced significant renewals in the practice of Krishna consciousness in order to make it universally applicable, he always emphasised the principle of 'simple living – high thinking' as an essential cultural approach, in which freshly established rural communities played a crucial role (Banyár, 2015). These communities were organised to minimise the effects of consumer society and to allow their members to concentrate on self-realisation and pursue an ecological lifestyle (Farkas, 2004). Currently, there are 48 such communities in 21 countries of the world. Krishna Valley is one of the 13 European ISKCON farm communities, which are situated in 12 countries (Kun, 2021).

After some humble beginnings in Communist Hungary in 1976, Krishna consciousness was registered as a religious organisation ('*egyház*' = 'church' in Hungarian) in 1989 thanks to the democratic transition and to its charismatic leader, Sivarama Swami (1949–), who left Hungary after the fall of the 1956 revolution and returned in the late 1980s as a Hare Krishna preacher (Kamarás, 1998). The legal adaptation of Krishna consciousness was not without challenges in the young Hungarian democracy, which was simultaneously characterised by the highly liberal and tolerant IV of 1990 *law on freedom of conscience and religion* as well as by a neo-conservative revival of the 'historic churches' exploited by right-wing political parties to suppress 'new religious movements'. Legally, neither the former nor the latter denomination was determined or prioritised. After a revision of approved religious organisations in 2011–2012, during which Krishna consciousness was suspended as a legally recognised religion along with Islam, Buddhism, or the Anglican Church for an upsetting six months, finally, it could maintain its status in the revised Law. Since religious censuses are uneven and unreliable in Hungary, it is the distribution of 1% of the personal income tax, which can be offered yearly to a religious organisation by every citizen – a system established by the 1990 Law –, which can be used as an indicator of the social acceptance of these organisations. According to the number of taxpayers offering 1% of their income tax to religious beneficiaries, since the early 2010s, HSKCON retains a solid fourth position after the three biggest Christian denominations (Roman Catholic, Calvinist, Lutheran) among the 32 recognised 'churches' (Banyár, 2021). It seems to be an obvious sign of the resilience of an organisation, which is not only one of the youngest among the approved confessions in Hungary but also was a target of political persecution in the early 1990s.

Amongst the ups and downs of the last three decades of Hungarian democracy, HSKCON could owe its relatively favourable situation to a series of conditions, which were duly and simultaneously mobilised by its leaders. The available secondary literature about HSKCON is rather scarce, thus, it was not sufficient to rely on its conclusions and we were conducting interviews with the leading figures of HSKCON to complete the available research results.[8] On the basis of the literature and the interviews, the following favourable conditions can be identified without determining their rank or their relative importance: (1) since the 16th century, Hungary has been a multi-confessional state, which has a long tradition of religious tolerance even though it was oppressed in different regimes. (2) Hungarian identity-building pre-national as well as modern national – always had a strong interest in its Asian/oriental origins, which makes the Hungarian society fairly open and curious about Eastern religions and philosophies. (3) HSKCON is part of the international network of ISKCON, which is supported by not only an extended system of devotees all over the world but also by the Indian state, which often regards it as its unofficial cultural ambassador. (4) In contrast with many religious organisations/churches, HSKCON could remain impartial in a society, which is excessively divided politically. (5) HSKCON could prove its societal use for this society through its extended charitable activities (food distribution for the needy, rehabilitation for young delinquents and drug addicts, hospice services, etc.). (6) HSKCON could profit from the high popularity of yoga by the fact that it, and especially its Bhaktivedanta College –the only state-accredited and – founded such higher education institution in Europe –, became flagship of yoga and yoga-related education in Hungary. (7) The care for the environment, ecological thought, and vegetarianism, which are integral components of Krishna consciousness and growingly appealing to young Hungarians also played a crucial part in the favourable reception of HSKCON. These conditions could not only ensure the resilience of this new religious organisation in Hungary, but they were also integrated into the establishment of Krishna Valley.

### 7.3.1   *Krishna Valley: resilience through participation*

Between 1993 and 2023, Krishna Valley went from an abandoned pastureland to the second most visited place in Somogy County[9] (Tóth-Soma, 2002). Out of Krishna Valley's annual 30,000 visitors, 10,000 are pilgrims and the rest are tourists who are interested in its multifaceted brand: Indian culture, yoga, organic farming and healthy living. A series of Gaudiya Vaishnava festivals is yearly celebrated with the participation of hundreds of devotees from several European countries. At the same time, there is also a yearly calendar for events inviting those who are interested in other aspects of Krishna Valley. Among these events, the most popular programme is the three-day Krishna Valley Fair since 1996, which, with its cultural performances, workshops, and educational presentations, attracts about 8,000 people each year, which is an impressive and challenging number to host for a community of 150 persons.

The seven elements, which were identified to understand the success of HSKCON is contemporary Hungary, can serve in the comprehension of Krishna Valley's development to this popular peripheral tourist destination. (1) HSKCON was barely known in the early 1990s in the Hungarian countryside when the land in the margins of Somogyvámos was purchased. The smooth integration of the Hare Krishna devotees was partially due to the mayor's attitude, who saw potential in the young settlers in the already bi-confessional (Roman Catholic, Lutheran) ageing village (Barabás, 1997). (2) Krishna Valley, although its origins have nothing to do with national ideologies, is associated with other rural initiatives, which strive for the revival of the Asian cultural roots of the Hungarians (Kocsis, 2004). One of such communities, the Kassai horseback archery valley, is only 30 kilometres away. (3) The growing Indian community as well as the Indian Embassy often consider Krishna Valley as its proper place for celebrating religious Hindu festivals. Moreover, there are several architectural inspirations (the shrine, the statues of war elephants at the main gate, etc.) from India, which determine the local landscape. Thus, it is not only that Krishna Valley's official designation is 'Indian cultural centre', but also provided the slogan of 'India's gateway in the heart of Europe'. (4) The community could always accomplish tight relations with the village hall regardless of its political composition due to its size (of potential voters within the village), its charitable activities, and to its favourable impact on the village's economy. (5) The community does not only provide regular food supply for the local needy, but it is also a model for rehabilitation nationwide. (6) Yoga camps and educational programmes co-organised by the Bhaktivedanta College enhance the events of the regular calendar and fill the empty periods. (7) The full designation of Krishna Valley is 'Indian cultural centre and ecological farm', which shows how significant ecological thought is in its identity. It is not only one of the largest and oldest ecovillages in Europe, but also a member of the Global Ecovillage Network of Europe, and its research institute has an observer status in the United Nations Framework Convention on Climate Change. Krishna Valley's representatives cooperate and share their experiences with a wide range of international organisations and higher educational institutions (Kun, 2021).

This very swift description of the multi-layered identity of Krishna Valley, which is simultaneously Krishna Valley and Eco-valley, shows an efficient governance model with a high level of adaptability. When the 280-hectare farm community was established by Sivarama Swami in compliance with the 'simple living – high thinking' principle, it was not necessarily meant to become today's showcase and tourist destination. Nevertheless, it could not have happened without the active participation and flexibility of the local community of devotees, who not only live in Krishna Valley anymore, but also formed a community in the nearby village (Farkas, 2009). It is important that Krishna Valley is not an open-air museum, but it remains a religious community for which it is also called '*New Vraja Dhāma*'. A – third – denomination evokes the birthplace of Lord Krishna, Vraja or Vrindavan in Northern India, which is the holiest place (*dhāma* in Sanskrit) for all Vaishnavas (Swami, 2012–2013). In this sense, this place also has a spiritual layer, which is displayed for the tourists as traditional Indian culture. The related tangible and intangible cultural goods (buildings, artefacts, clothing, instruments, rituals, songs, etc.) and activities (cow protection, yoga, vegetarian cooking, etc.)

can be concomitantly interpreted as spiritual, Indian (i.e., exotic in Europe), and ecological. It is probably not surprising that the latest initiative of the local leaders[10] is to unite these layers under the denomination of 'heritage' revealing the very high flexibility of the third regime cultural heritage uniting tangible, intangible, and natural aspects of a locality. As in the Újpest case, here also, the resilient governance lies in the creative aggregation of cooperation, adaptability, and successfully exhibited self-representation.

## 7.4    Conclusion

This chapter reflected on resilient heritage communities and the potential of community-based, sustainable cultural tourism in the case of two peripheral – urban and rural – Hungarian areas. The challenges and opportunities of cultural tourism and the explicit involvement of local communities could be detected through innovative participatory approaches and practices. Whether we are analysing the Újpest case which stands out as an eminent example of minority heritage preservation or the Krishna Valley which can be observed as a model of sustainable-ecological, but also spiritual and religious heritage places, the third heritage regime seemed as a pertinent approach. Its renewed perspective, emphasising community heritage, cultural rights, and intangibility is an adequate viewpoint and possibility of analysis to uncover hidden potentials in remote, marginal areas.

### Notes

1 Romani Studies Program (https://www.ceu.edu/unit/romani-studies-program – May 13, 2020).
2 European Roma Institute for Arts and Culture (https://eriac.org/about-eriac/ – May 13, 2020).
3 Critical Romani Studies (https://crs.ceu.edu/index.php/crs – May 13, 2020).
4 Interview with István Gábor Molnár, the president of Roma local government of Újpest, on 10 April 2019 conducted by Eszter György.
5 An article written by István Gábor Molnár about the history of the urban neighbourhood: https://kisebbsegkutato.tk.hu/uploads/files/olvasoszoba/romaszovegtar/Ujpesti_ciganysag_tortenete.pdf.
6 As the article proves, the results of this research are manifold, but besides the above-mentioned exhibitions and other facilities, the history of the local Roma community was also published and disseminated by István Gábor Molnár several times. Among others, an article entitled: Settlers attached to their local area? The history of the Roma in Újpest https://epa.oszk.hu/03900/03995/00057/pdf/EPA03995_naput_2010_07_019-041.pdf, A family photo collection https://rofodia.oronk.hu/fenykeptar/ujpesti-csaladi-fotok-molnar-istvan-gabor-gyujtemenye/, and a video presentation in the frame of Romakép Műhely (Roma Visual Lab) https://www.youtube.com/watch?v=XRrS-68MbY8&ab_channel=Romak%C3%A9pM%C5%B1hely.
7 Újpest Roma Local History Collection and Gyöngyi Rácz Community Center (https://www.facebook.com/ujpesticigany – 6 February 2024).
8 Five interviews were made in 2023 with the leaders of this religious organisation and their anonymity is respected.

9  The most attractive tourist destination is Lake Balaton.
10  In 2023, legal actions were inducted to unite the different legal entities related to Krishna-valley under a unique foundation entitled 'For the preservation of Created Heritage'.

## Bibliography

Acton, T. (1974). *Gypsy Politics and Social Change: The Development of Ethnic Ideology and Pressure Politics among British Gypsies from Victorian Reformism to Romany Nationalism*. Routledge and K. Paul.

Allen, D. (2018). Resilient Cultural Transmission in a Hostile State. In D. Allen, M. Greenfields, & D. Smith (Eds.), *Transnational Resilience and Change: Gypsy, Roma and Traveller Strategies of Survival and Adaptation* (pp. 9–40). Cambridge Scholars Publishing.

Baar, H. J. M. van. (2011). *The European Roma: Minority Representation, Memory, and the Limits of Transnational Governmentality*. Universiteit van Amsterdam.

Banyár, M. (2015). A Kṛṣṇa-tudat mint egyházi intézmény. [Krishna-consciousness as a religious institution.] *Tattva*, *XVIII*, 79–100.

Banyár, M. (Maharani Devi Dasi) (2021). *Kṛṣṇa társadalma*. [Krishna's society.] Lāl.

Barabás, M. (1997). *Közösségek találkozása: Krisna-völgy Somogyvámoson*. [Meeting communities: Krishna-valley in Somogyvámos.] MTA PTI Etnoregionális Kutatóközpont.

Beel, D. E., Wallace, C. D., Webster, G., Nguyen, H., Tait, E., Macleod, M., & Mellish, C. (2017). Cultural Resilience: The Production of Rural Community Heritage, Digital Archives and the Role of Volunteers. *Journal of Rural Studies*, *54*, 459–468. https://doi.org/10.1016/j.jrurstud.2015.05.002

Bendix, R., Eggert, A., & Peselmann, A. (2017). *Heritage Regimes and the State*. Universitätsverlag Göttingen.

Bollig, M. (2014). Resilience – Analytical Tool, Bridging Concept or Development Goal? Anthropological Perspectives on the Use of a Border Object. *Zeitschrift Für Ethnologie*, *139*(2), 253–279.

Böschen, S., Binder, C. R., & Rathgeber, A. (2019). Resilience Constructions: How to Make the Differences between Theoretical Concepts Visible? In B. Rampp, M. Endreß, & M. Naumann (Eds.), *Resilience in Social, Cultural and Political Spheres* (pp. 11–39). Springer Fachmedien Wiesbaden. https://doi.org/10.1007/978-3-658-15329-8_2

Bowers, J. (2009). Gypsies and Travellers Accessing their Own Past: The Surrey Project and Aspects of Minority Representation. In M. Hayes & T. Acton (Eds.), *Travellers, Gypsies, Roma: The Demonisation of Difference* (pp. 17–29). Cambridge Scholars Publishing.

Brand, F. S., & Jax, K. (2007). Focusing the Meaning(s) of Resilience: Resilience as a Descriptive Concept and a Boundary Object. *Ecology and Society*, *12*(1). https://doi.org/10.5751/ES-02029-120123

Chandler, D. (2014). Beyond Neoliberalism: Resilience, the New Art of Governing Complexity. *Resilience*, *2*(1), 47–63. https://doi.org/10.1080/21693293.2013.878544

Endreß, M. (2019). The Socio-Historical Constructiveness of Resilience. In B. Rampp, M. Endreß, & M. Naumann (Eds.), *Resilience in Social, Cultural and Political Spheres* (pp. 41–58). Springer Fachmedien Wiesbaden. https://doi.org/10.1007/978-3-658-15329-8_3

Farkas, J. (2004). "Minden lépés tánc, minden szó egy ének": A mindennapi élet valósága a Krisna-tudatban. ['Every step is a dance, every word is a song': Realities of everyday life in Krishna consciousness]. In É. Pócs (Ed.), *Múlt és jelen: Tudományos konferencia a Pécsi Tudományegyetem Néprajz Tanszékének 10 éves jubileumán 2001. Szeptember 17–18-án* (pp. 151–181). L'Harmattan.

Farkas, J. (2009). Ardzsuna *dilemmája: reszocializáció és legitimáció egy magyar Krisna-hitű közösségben.* [Arjuna's dilemma: resocialisation and legitimation in a Hungarian Krishna-conscious community.] L'Harmattan.

Greenfields, M., & Smith, D. (2018). Resisting Assimilation: Survival and Adaptation to 'Alien' Accommodation Forms. The Case of British Gypsy/Travellers in Housing. In D. Allen, M. Greenfields, & D. Smith (Eds.), *Transnational Resilience and Change: Gypsy Roma and Traveller Strategies of Survival and Adaptation* (pp. 94–114). Cambridge Scholars Publishing.

György, E., & Oláh, G. (2018). The Creation of Resilient Roma Cultural Heritage. Case Study of a Bottom-Up Initiative from North-Eastern Hungary. *Socio.Hu,* *8*(Special Issue 6), 37–50. https://doi.org/10.18030/socio.hu.2018en.31

Hartog, F. (2015). *Regimes of Historicity: Presentism and Experiences of Time.* Columbia University Press. https://doi.org/10.7312/hart16376

Humbert, C., & Joseph, J. (2019). Introduction: The Politics of Resilience: Problematising Current Approaches. *Resilience, 7*(3), 215–223. https://doi.org/10.1080/216 93293.2019.1613738

Joseph, J. (2013). Resilience as Embedded Neoliberalism: A Governmentality Approach. *Resilience, 1*(1), 38–52. https://doi.org/10.1080/21693293.2013.76574 1Kamarás, I. (1998). *Krisnások Magyarországon.* [Krishna-believers in Hungary.] Iskolakultúra.

Kocsis, G. (2004). Adalékok a Krisna-tudat és a magyar társadalom kapcsolatához. [Thoughts about the relationship between Krishna-consciousness and the Hungarian society.]. In A. A. Gergely & R. Papp (Eds.), *Kisebbség és kultúra* (pp. 274–298). MTA Etnikai-nemzeti Kisebbségkutató Intézet - MTA Politikai Tudományok Intézete-ELTE Kulturális Antropológiai Szakcsoport.

Kun, A. (Ed.) (2021). *Beszélgetések az önellátásról 2.: a biogazdálkodás és az ökologikus életmód alapjai.* [Conversations about self-sufficiency 2: the basics of organic farming and ecological lifestyle] Öko-völgy Alapítvány.

Lazzeretti, L. (2012). The Resurge of the "Societal Function of Cultural Heritage". An Introduction. *City, Culture and Society, 3*(4), 229–233. https://doi.org/10.1016/j. ccs.2012.12.003

Mckeown, A., & Glenn, J. (2018). The Rise of Resilience after the Financial Crises: A Case of Neoliberalism Rebooted? *Review of International Studies, 44*(2), 193–214. https://doi.org/10.1017/S0260210517000493

Meyen, M., & Schier, J. (2019). The Resilience Discourse: How a Concept from Ecology Could Overcome the Boundaries between Academic Disciplines and Society. In B. Rampp, M. Endreß, & M. Naumann (Eds.), *Resilience in Social, Cultural and Political Spheres* (pp. 105–120). Springer Fachmedien Wiesbaden. https://doi. org/10.1007/978-3-658-15329-8_6

Nasture, F. (2015). Changing the Paradigm of Roma inclusion: From Gypsy Industry to Active Citizenship. In K. O'Reilly & M. Szilvasi (Eds.), *Roma Rights 2 2015: Nothing about Us Without Us? Roma Participation in Policy Making and Knowledge Production* (pp. 27–31). European Roma Rights Centre.

Obrist, B., Pfeiffer, C., & Henley, R. (2010). Multi-Layered Social Resilience: A New Approach in Mitigation Research. *Progress in Development Studies, 10*(4), 283–293. https://doi.org/10.1177/146499340901000402

Okely, J. (2012). Constructing Culture Through Shared Location, Bricolage and Exchange: The Case of Gypsies and Roma. In M. Stewart & M. Rövid (Eds.), *Multi-Disciplinary Approaches to Romany Studies* (pp. 196–210). Central European University.

Rampp, B. (2019). The Question of 'Identity' in Resilience Research. Considerations from a Sociological Point of View. In B. Rampp, M. Endreß, & M. Naumann (Eds.), *Resilience in Social, Cultural and Political Spheres* (pp. 59–76). Springer Fachmedien Wiesbaden. https://doi.org/10.1007/978-3-658-15329-8_4

Rostas, I., & Rövid, M. (2015). On Roma Civil Society, Roma Inclusion, and Roma Participation. *Roma Rights,* No 2, 7–10.

Ruethers, M. (2013). Jewish Spaces and Gypsy Spaces in the Cultural Topographies of a New Europe: Heritage Re-Enactment as Political Folklore. *European Review of History: Revue Européenne d'histoire, 20*(4), 671–695. https://doi.org/10.1080/13507486.2013.809567

Scott, J. C. (2009). *The Art of Not Being Governed: An Anarchist History of Upland Southeast Asia*. Yale University Press.

Silverman, C. (2014). *Romani Routes: Cultural Politics and Balkan Music in Diaspora*. Oxford University Press.

Smith, L. (2010). *Uses of Heritage*. Routledge.

Sonkoly, G. (2017). *Historical Urban Landscape*. Springer International Publishing. https://doi.org/10.1007/978-3-319-49166-0

Sonkoly, G., & Vahtikari, T. (2018). *Innovation in Cultural Heritage Research: For an Integrated European Research Policy*. European Commission.

Stewart, M. (1994). *Daltestvérek: Az oláhcigány identitás és közösség továbbélése a szocialista Magyarországon* [Brothers in song: The persistence of (Vlach) Gypsy identity and community in Socialist Hungary] (Vol. 2). Magyar Tudományos Akadémia Szociológiai Intézet.

Swami, S. (2012–2013). *Nava-vraja-mahimā* (Vol. I–VIII). Lāl.

Tóth-Soma, L. (Ed.) (1996). *Hare Krisna fehéren-feketén* [Hare Krishna in black and white] Agóra Print.

Tóth-Soma, L. (Ed.) (2002). *Kérdések és válaszok a Krisna-tudatról* [Questions and answers about Krishna-consciousness]. Lāl.

Trehan, N., & Kóczé, A. (2009). Racism, (Neo-)Colonialism and Social Justice: The Struggle for the Soul of the Romani Movement in Post-Socialist Europe. In G. Huggan & I. Law (Eds.), *Racism Postcolonialism Europe* (pp. 50–74). Liverpool University Press. https://doi.org/10.5949/liverpool/9781846312199.003.0004

Walker, J., & Cooper, M. (2011). Genealogies of Resilience: From Systems Ecology to the Political Economy of Crisis Adaptation. *Security Dialogue, 42*(2), 143–160. https://doi.org/10.1177/0967010611399616

Wilson, G. A. (2017). Resilience and Human Geography. In D. Richardson, N. Castree, M. F. Goodchild, A. Kobayashi, W. Liu, & R. A. Marston (Eds.), *International Encyclopedia of Geography: People, the Earth, Environment and Technology* (pp. 1–10). John Wiley & Sons, Ltd. https://doi.org/10.1002/9781118786352.wbieg1205

# 8 Participatory platform for sustainable cultural tourism

## The case of Central Slovakia

*Darina Rojíková, Kamila Borseková
and Alexandra Bitušíková*

### 8.1 Introduction

This chapter aims to present and evaluate the participatory methodology employed in the development of "Mining Treasures," an interactive online platform designed to enhance regional development and tourism in the Banska Bystrica Self-Governing Region of Slovakia. This platform, particularly focused on the region's less developed and marginalised areas, was created through extensive collaboration with diverse stakeholders and individuals. The underlying concept of this initiative is sustainable cultural tourism, which, according to the European Commission (2019), involves the integrated management of cultural heritage and tourism activities in harmony with local communities. This approach not only aims to yield social, environmental, and economic benefits for all involved parties but also emphasises the conservation of both tangible and intangible cultural heritage alongside sustainable tourism development. Sustainable cultural tourism places the cultural heritage and its communities at the heart of decision-making, ensuring that the benefits of tourism are shared with the cultural heritage and its local people. This concept supports the preservation of cultural heritage and its authentic interpretation, as well as local sustainable economies. Furthermore, sustainable cultural tourism seeks to balance the benefits of attracting visitors to local communities with the mitigation of potential negative impacts, such as the degradation of cultural sites and practices due to overuse and commodification. The chapter will also explore strategic planning and networking in sustainable cultural tourism, highlighting concepts like "slow tourism," "authenticity," "storytelling," "well-being," and "contact with locals" (Callot, 2013). The European Commission (2019, 2022) underscores the integral role of society, heritage communities, groups, and individuals in cultural heritage, not merely as passive audiences but as active participants in governance and management (for more information see, e.g. Borseková et al., 2022, 2023).

Therefore, the primary motivation behind this chapter is to share our experiences with a broader professional audience regarding the participatory creation of a responsive web platform and digital map, "Mining Treasures of Central Slovakia." This platform vividly showcases the paramount mining

DOI: 10.4324/9781003422952-8

locations in the region's two cities, Banská Bystrica and Banská Štiavnica (and their environs). It not only presents these sites in an engaging manner but also fosters participation and heritage community building. The participatory approach adopted during its development aims to rejuvenate the current tourism offerings in the market. It seeks to distribute tourists more evenly, especially diverting them from crowded cities to broader areas, attract those keen on mining cultural heritage (be it tangible, intangible, or industrial), and provide unparalleled educational resources tied to the region's mining history and culture. This is particularly crucial given the region's underutilisation of its cultural heritage for sustainable tourism and territorial development. While cities grapple with the challenges of mass tourism, often drawing criticism from locals, the surrounding rural municipalities face a dearth of tourist interest and inadequate development. Hence, a key goal of the platform is to distribute tourists more evenly across a larger area that has much to offer.

The potential of the Banská Bystrica region is deeply intertwined with its mining legacy. Since the 13th century, deposits of gold, silver, and copper, especially in Banská Bystrica, Banská Štiavnica, and Kremnica, have been exploited. Reflecting the metals mined, the mining towns were aptly named "Copper Banská Bystrica," "Silver Banská Štiavnica," and "Golden Kremnica."

The city of Banská Bystrica, known for copper mining, attained its current picturesque appearance in the late Middle Ages. This transformation was largely due to the affluent Fugger and Thurzo families, who established the prosperous Thurzo-Fugger Copper Company in 1494, a pioneering capitalist enterprise of its time. The Thurzo-Fugger company is also credited with several global "firsts," including the introduction of an 8-hour workday, retirement benefits for miners and widows, medical care, and the innovative double-entry bookkeeping system. The region is replete with over 250 tangible cultural heritage sites, including palaces, mansions, and more, particularly in and around Banská Bystrica.

Another significant area with a rich mining history is Banská Štiavnica, known as "Silver Banská Štiavnica." This historic city, along with its surrounding technical monuments, has been a UNESCO World Heritage site since 1993. Banská Štiavnica has a storied past, from its early mentions in 1156 as "terra banensium" to its rise as the largest mining centre in the Habsburg monarchy in the 18th century. The city is home to 360 monuments, with over 200 linked to its mining history.

Despite the immense potential of this heritage for tourism, its value remains underappreciated in the region. There's a noticeable absence of marketing and digital tools that could spotlight this unique historical facet, fostering territorial development and sustainable tourism.

## 8.2   The current state of tourism

The Banská Bystrica Self-Governing Region, spanning an area of 9,454 km² and constituting 19.3% of Slovakia's total land area, is the country's largest region. Located in the southern part of central Slovakia, it shares its southern border with the Republic of Hungary. The region's topography is diverse, ranging from high mountainous terrains in the north, rugged landscapes in the centre, to gently rolling and flat areas in the south.

With a population of approximately 660,000, the Banská Bystrica Self-Governing Region has a population density of fewer than 70 inhabitants per km². Once considered the most prosperous and affluent region in Slovakia, it now ranks among the least developed. This decline is evident from its relatively high unemployment rate and low Gross Domestic Product (GDP). Despite the region's rich cultural and natural assets, a significant portion of its population is employed in the industrial, trade, and construction sectors. Mining, once a pivotal industry for the region, has seen a decline, with the focus now primarily on surface mining of non-metallic minerals. The industrial sector contributes to over a quarter of the region's gross added value. In contrast, trade, transport, accommodation, and catering—sectors intertwined with tourism—contribute only 17% to the added value.

Relative to other Slovak regions, the tourism sector in the Banská Bystrica Self-Governing Region is underdeveloped. This observation is supported by tourism metrics. Tourism revenues in the region account for a mere 0.57% of Slovakia's total, the lowest among all Slovak regions. This trend is consistent across various tourism indicators, including revenues from active tourism (2.26%), passive tourism (0.44%), domestic tourism (2.85%), and foreign tourism (1.36%).

One significant factor impacting the attractiveness of this region is the accessibility of transportation infrastructure modes such as trains, buses, and roads. Despite the fact that the region has a network of public bus and train transport, this is not sufficiently integrated. There is also a lack of road connection between the east-west and north-south of Slovakia, which would quickly and qualitatively connect the Banská Bystrica region with other touristic significant locations in Slovakia with a high concentration of tourists, e.g. the High Tatras, the capital Bratislava, Košice. In some cities of the region, e.g. Banská Bystrica and Banská Štiavnica, there is also a huge problem with parking. There is a lack of parking lots with the possibility of public transportation to city centres. The lack of tourist information centres with multilingual staff and insufficient promotion of available services in the region contribute to a perception of poor infrastructure and detract from the overall tourist experience.

Some of the mentioned problems that create barriers for the development of tourism in the Banská Bystrica region are the subject of discussions by politicians, experts, and the general public. Currently, there are slowly emerging solutions that could remove these barriers (e. g. integrated public transport system, completion of important roads connecting the southern and northern parts of the region). However, their implementation and the resulting effect will not be seen until several years from now.

### 8.2.1    *Indicators of tourist potential*

From a tourism perspective, there are notable disparities between the districts of Banská Bystrica and Banská Štiavnica, largely attributed to their differing sizes. The Banská Bystrica district houses up to a quarter of the region's accommodation facilities. However, when considering the distinct sizes of the districts and the cities themselves, the difference in the number of accommodations—70 in both Banská Bystrica and Banská Štiavnica—is relatively minor.

The most significant variances between the two districts emerge when examining indicators like the total number of rooms and beds. The Banská Bystrica district offers over a third of the region's total accommodation capacity, but only a quarter of this capacity is utilised in both districts. In contrast, the Banská Štiavnica district primarily features smaller accommodation facilities, as evidenced by the room and bed counts per facility.

Although the Banská Bystrica district encompasses a third of the region's accommodation capacities, it accounts for 73% of the total accommodation revenue. This trend is consistent with other indicators, such as total revenue from both foreign and domestic visitors.

The reason is that Banská Bystrica district also attracts a significant number of business travellers due to its central location, developed infrastructure, presence of industries, businesses, and government institutions. The district's conference facilities, hotels, and amenities cater to the needs of corporate travellers, who often spend more on accommodation and related services compared to leisure tourists. Banská Bystrica district also serves as a hub for economic activities and employment opportunities in the region. Overall, the combination of appealing tourist attractions, business opportunities, infrastructure, and economic activities contributes to Banská Bystrica district's dominance in accommodation revenues within the region. This phenomenon underscores the district's importance as a tourism and economic hotspot and highlights the potential for further growth and development.

Despite the average accommodation prices in both districts being below the regional mean, domestic visitors often express concerns about the costs. The high demand for tourism services, especially in Banská Štiavnica, has led to an increase in prices for related services, like dining, over recent years. This has made them less affordable, particularly for residents.

Moreover, essential service providers like hair salons, grocery stores, clothing retailers, and drugstores have gradually vanished from Banská Štiavnica's centre. In their place, tourism-related establishments like cafes, bakeries, restaurants, and souvenir shops have emerged. This shift has posed accessibility challenges, especially for locals residing in the city centre. Such developments, among others, have fostered tension between the local community and tourists.

The trends in the indicators "Overnight Stays in Accommodation Facilities" (Figure 8.1) and "Turnover in Accommodation Facilities" (Figure 8.2)

*Table 8.1* Indicators of tourist potential in the Banská Bystrica region, Banská Bystrica, and Banská Štiavnica districts

| Indicator of tourism | Banská Bystrica region | Banská Bystrica district | | Banská Štiavnica district | |
| --- | --- | --- | --- | --- | --- |
| Total number of accommodation facilities available | 61,400 | 15,600 | 25% | 8,600 | 14% |
| Total number of rooms available | 7 97,300 | 2 91,100 | 37% | 57,600 | 7% |
| The number of rooms converted to 1 accommodation facility | 13 | 19 | | 7 | |
| Total number of beds (including camping places) available | 2,1 47,000 | 7 48,200 | 35% | 1 76,600 | 8% |
| The number of beds converted to 1 accommodation facility | 35 | 47 | | 20 | |
| Total revenue for accommodation (euro) | 1,2 37,2 13,000 | 9 04,0 70,300 | 73% | 2 29,6 25,100 | 19% |
| Sales of foreign visitors (EUR) | 2 24,2 99,900 | 1 56,7 28,900 | 70% | 327 633,00 | 15% |
| Sales of domestic visitors (euro) | 1,0 80,4 84,100 | 6 79,7 70,400 | 63% | 1 96,8 61,800 | 18% |
| Average price for accommodation in accommodation facilities (euro) | 3,670 | 3,130 | | 3,020 | |
| Average price for accommodation in accommodation facilities per foreign visitor (euro) | 5,050 | 3,440 | | 3,730 | |
| Average price for accommodation in accommodation facilities per domestic visitor (euro) | 3,530 | 3,040 | | 2,930 | |
| Net use of permanent beds in accommodation facilities (%) | 2,490 | 2,360 | | 2,220 | |
| Net room utilisation (%) | 3,330 | 2,920 | | 2,680 | |

*Source:* Own elaboration based on data from Statistical Office of the Slovak Republic

highlight the seasonality of tourism in the districts of Banská Bystrica and Banská Štiavnica. Notable differences between the districts are evident.

In the Banská Bystrica district, there are two primary tourist seasons: summer and winter. However, each lasts only for a brief duration of two months. The summer season, occurring in July and August, aligns with Slovakia's summer holidays. The winter season spans from January to February.

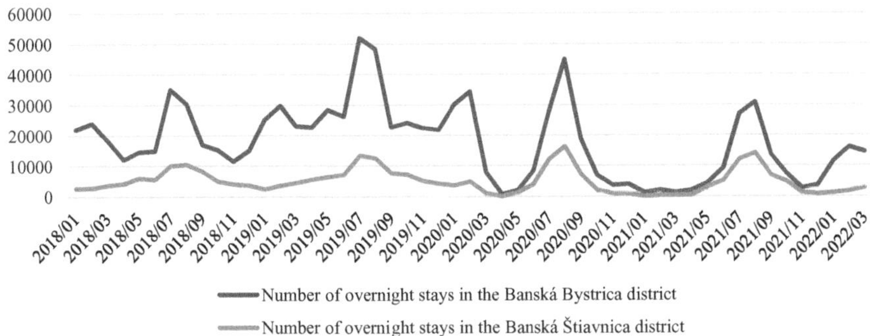



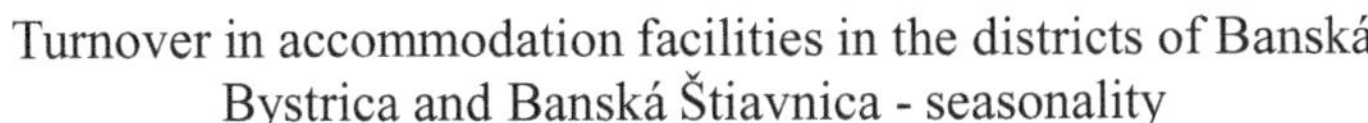



*Figure 8.1* Overnight stay in accommodation facilities in the districts of Banská Bystrica and Banská Štiavnica.

*Source:* Own elaboration based on data from Statistical Office of the Slovak Republic

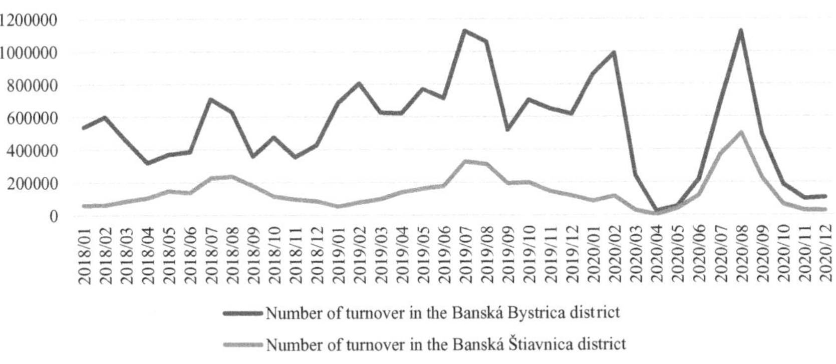





*Figure 8.2* Turnover in accommodation facilities in the districts of Banská Bystrica and Banská Štiavnica.

*Source:* Own elaboration based on data from Statistical Office of the Slovak Republic

Conversely, the Banská Štiavnica district has a single, more extended season that gradually builds up and lasts for four months, with peak tourist activity observed from June to September.

The distinct seasonality between the two districts can be attributed to their unique offerings. Banská Bystrica, with several winter resorts in its vicinity, is better equipped for winter tourism. On the other hand, Banská Štiavnica boasts a rich cultural heritage linked to mining and is home to 24 artificial lakes (from an original count of 60). These lakes, initially designed innovatively to generate energy for mines, now serve as recreational spots during the summer.

The stretch from April to May 2020 and December to April 2021 was particularly unique for tourism in both districts due to the challenges posed by the COVID-19 pandemic and the associated preventive measures.

### 8.2.2  *Identification of competitive advantage*

Banská Bystrica and Banská Štiavnica boast a remarkable mining history dating back centuries. Banská Štiavnica and Banská Bystrica were once among the most important mining towns in the Kingdom of Hungary. Their territories showcase well-preserved mining structures, including mines, churches, and manor houses. Banská Bystrica and Banská Štiavnica, along with their surrounding municipalities, are distinguished by over 600 tangible and industrial heritage sites that stand testament to their rich mining history. The cultural vibrancy of Banská Bystrica and Banská Štiavnica is evident through numerous festivals and events celebrating their heritage. Additionally, region hosts annual cultural events, such as historical reenactments, folklore festivals, and arts exhibitions, which attract visitors from across Slovakia and beyond. This cultural heritage, intertwined with the region's stunning natural landscapes and punctuated by cultural routes like the Barbora and Fugger routes, establishes Central Slovakia's unique competitive advantage.

While other regions of Slovakia may possess historical sites and cultural attractions, few rival the depth, diversity, and authenticity found in Banská Bystrica and Banská Štiavnica. By harnessing and effectively leveraging this shared competitive advantage, mining towns and tourist routes can significantly bolster the region's sustainable development. This aligns with the findings of Borseková et al. (2015, 2017), which highlight that the most commonly identified competitive advantage across all Slovak regions is their cultural and historical heritage, serving as the primary tourist attraction in Slovakia. Such a heritage not only offers a type of competitive advantage but also holds the potential to evolve into a genuine competitive advantage, given its unique and inimitable nature.

### 8.2.3  *Routes connecting the mining cultural heritage in Central Slovakia*

The **Barbora Route** stands as the longest educational and sightseeing route in Slovakia, weaving together mining monuments and attractions throughout central Slovakia. Strategically segmented between villages, the route can be traversed on foot or by bicycle. After completing each segment, visitors have the option to rest at local accommodation facilities, which also offer luggage transfer services. Along the way, catering services present an opportunity for travellers to taste traditional local dishes.

Spanning 186.2 km, this pilgrimage route begins and concludes in Banská Bystrica, designed to be completed over nine days. It's structured into nine stages, with each stage culminating in an overnight stay. The stages vary in length, alternating between longer and shorter distances. As the name suggests, the route pays homage to St. Barbora, the patron saint of miners. It celebrates locations tied to the mining history of both Slovak and broader European significance. The Barbora route features 29 symbolic stops, commemorating the age at which Saint Barbora passed away. These stops highlight locations of mining, historical, natural, religious, cultural, and technical significance. Throughout the journey, visitors can collect stamps in a guidebook, and upon completion with all stamps, they receive a certificate of passage.

The **Fugger Route** serves as an educational exploration trail leading to mines and mountain centres that enriched the Fuggers with silver, copper, and iron. The Fuggers amassed their wealth in early modern Augsburg through the cotton trade, loans to prominent figures, and predominantly the mining industry. Circa 1490, the Fuggers established a pan-European mining enterprise with ore mines and smelters in Tyrol, Carinthia, and Banská Bystrica in Upper Hungary (present-day Slovakia). Their ventures with various metals, especially copper and mercury, penned a riveting chapter in European economic and social history. Today, Fugger houses, castles, churches, monuments, accessible mining tunnels, and museums in cities like Banská Bystrica, Kremnica, and Banská Štiavnica offer insights into the history of this mining conglomerate, which, by around 1660, prefigured the European Union.

Today's Fugger Route in Slovakia is a segment of the broader European Fugger Route, linking cities such as Augsburg, Bad Hindelang, Schwaz, Hall, Sterzing, and Banská Bystrica. Banská Bystrica (Neusohl) stands as the heart of the Fugger Route in present-day Slovakia, given the Fuggers' significant copper ore acquisitions from the city. An old adage goes, "Golden Augsburg stands on Copper Banská Bystrica."

In contrast to the Barbora Route, which is actively marketed as a tourism product, the Fugger Route lacks management and promotion. The Barbora Route's stewardship lies with the civic association Terra Montanae. Its members maintain the hiking trails, mark the route, and guide pilgrims along the way. In the realm of sustainable tourism, a participatory approach stands as a cornerstone for genuine and lasting development. By actively involving local communities, stakeholders, and visitors in decision-making processes, we can ensure that tourism initiatives are both environmentally sustainable and culturally respectful. The Barbora and Fugger Routes, with their rich historical and cultural significance, present a prime opportunity for such development. By harnessing wider participation from local communities and beyond, these routes can be transformed into vibrant hubs of sustainable tourism, ensuring their preservation and relevance for generations to come.

## 8.3    Participatory platform—Mining treasures of Central Slovakia

The responsive web platform and digital map, "Mining Treasures of Central Slovakia," vividly showcase the most significant mining sites in two key cities of the region: Banská Bystrica and Banská Štiavnica, as well as their surrounding areas. The participatory approach adopted in its development not only refreshes the current tourism offerings but also helps disperse tourists, especially those from urban areas, across a broader region.

Informing visitors about the offerings of the region beyond the city centre is important, but it is not sufficient on its own to redirect tourist flows. The Mining Treasures of Central Slovakia platform adopts a multifaceted approach that combines the promotion of both well-known and lesser-known mining sites, education, and cooperation with local stakeholders.

The Mining Treasures of Central Slovakia platform employs a comprehensive and strategic approach to marketing activities aimed at promoting mining cultural heritage, thus mitigating the effects of tourism on city centres, especially in Banská Štiavnica. The platform not only highlights renowned sites but also showcases lesser-known mining locations. This encompasses the promotion of industrial and material heritage, as well as cultural events and tourism in the surrounding villages. Lesser-known mining sites often remain undiscovered by tourists due to the lack of presentation and promotion by tourist information centres, tourism organisations, and travel agencies. Additionally, the platform offers off-season activities through tourism and cultural events. By diversifying tourism offerings, the platform can distribute visitor flows and alleviate pressure on city centres.

However, for the implementation of sustainable tourism practices, cooperation with local stakeholders such as entrepreneurs, civic associations, cultural institutions, and residents, who acknowledge the adverse impacts of tourism on overcrowded city centres, is crucial. These stakeholders actively contribute to the platform's content creation, thereby taking ownership and spreading its message further. By highlighting the unique offerings of the entire region, not just the city centre, stakeholders advocate for responsible tourism. The platform's educational function, featuring distinctive educational resources that delve into the region's mining history and culture, also plays a vital role in promoting sustainable tourism in the area. It is essential for children and young people to learn about the region's rich and rare mining history, fostering their readiness to protect cultural heritage and engage in sustainable and responsible tourism in the future.

The platform also provides a forum for stakeholders to engage in discussions with representatives of local and regional governments regarding potential solutions for promoting sustainable tourism in the region. Government investment in infrastructure beyond the city centre is crucial, including improvements to transport links, signage, and visitor facilities. These enhancements can facilitate tourists' access to and enjoyment of attractions outside the city centre.

Through these measures, the platform can assist in alleviating the adverse impacts of tourism on city centres, fostering a more sustainable and equitable tourism experience for visitors, and generating or sustaining tourism income for residents.

This approach attracts a diverse set of tourists, especially those keen on exploring mining cultural heritage—be it tangible, intangible, or industrial.

The participatory creation of the "Mining Treasures of Central Slovakia" platform commenced in June 2021 and was officially launched in December 2022 on the website www.banickepoklady.eu. As of now, it features a total of 250 activities spread across nine categories: galleries and museums; tangible heritage—monuments; intangible heritage; industrial heritage; events; education; for children; hiking; and on the map. "Mining Treasures" is a dynamic platform, regularly updated with new activities, intriguing features, and news articles. Since April 2023, an English version has been available, broadening its appeal to international visitors. Since its inception, the "Mining Treasures of Central Slovakia" platform has garnered 338 users from 18 different countries, with an average visit duration of 2 minutes and 7 seconds. The majority of users are from Slovakia, followed by the USA, Sweden, and Ireland. On the platform, the cities of Banská Bystrica and Banská Štiavnica, articles, and routes—specifically the Barbora and Fugger routes—as well as activities related to galleries and museums have been of particular interest to visitors. The platform's promotion is facilitated through its integration with social media platforms like Facebook, Instagram, and YouTube. While "Mining Treasures of Central Slovakia" is fundamentally a digital platform, its creation extends beyond just the technical aspects. The participatory development process encompassed data collection, community engagement, design, marketing, and content creation.

### 8.3.1  *Data collection*

As part of the creation of the platform "Mining Treasures of Central Slovakia," primary and secondary data related to the creation, testing, use and promotion of the interactive platform were collected and evaluated. Data related to the creation of the interactive platform consists of primary and secondary data related to the content of the interactive platform; this includes texts, photographs, videos, and audio materials. Data related to the testing, use, and promotion of the interactive platform consists of collecting data related to the testing and usage of the platform, such as the number of visits/ clicks, the number of interactions, and data related to feedback while using the interactive platform (from local communities and beyond) and collecting data related to using/promoting the platform and its content through social media (e.g., using the hashtag #banickepoklady/#miningtreasures, liking, commenting, sharing the content of the platform through different types of social media).

In connection with data collection for interactive platform "Mining Treasures of Central Slovakia," more than 30 museums and galleries, 60 tangible heritage objects, 40 industrial heritages objects and 25 events were physically visited. More than 250 texts were prepared and reviewed for the content of the interactive platform, more than 2,300 photos were taken, and more than 350 websites were visited to get inspired. A total of more than 100 people participated in data collection within the "Mining Treasures of Central Slovakia" platform.

### 8.3.2 Meetings with communities

At the very beginning of the creation of the interactive platform "Mining Treasures of Central Slovakia," several important meetings were held with key stakeholders. This includes the Regional Destination Management Organisation BBSK and the Local Destination Management Organisation Central Slovakian Local Destination Management Organisation. The participation of students who became active contributors to the platform by creating original content for the platform was increased. In addition, more than 20 meetings were organised with the ICT company related to the preparation phase of the interactive digital platform. Training on the functionality of the content of the interactive platform and adding items was also carried out by the company that designs the interactive platform.

Through destination management organisations and numerous meetings with local organisations, new groups of local communities and individuals are involved in the creation of the platform. Successfully involved in the creation of the platform are local photographers who provide photos and visual materials for the content of the platform.

University students are involved in the creation of content as well as testing and promotion of the platform through specific assignments in five different courses (Creativity and culture in regional development; New trends in local and regional politics; Participatory creation of public policy; Basics of marketing; and Marketing public and non-profit sector). Students' participation led to the creation of 19 proposals for the logo and design manual of the interactive mining treasure platform. The winning proposal is used in the final version of the platform. Students also become active contributors and users of the platform. At the beginning of 2023, we met the author of the children's book "*Copper Land*," which tells the story of the Špania Dolina treasure and how a forest elf became a permonian. We were also invited to the launch of the book and established cooperation with this author. In 2023, it was also possible to meet and develop cooperation with stakeholders who are significantly involved in the protection and development of the mining heritage for the development of sustainable tourism and the creation of new products, including CA Terra Montanae, CA Štiavnický tajch, Primary school with kindergarten Maxilian Hella Štiavnické Bane, CA Špaňodolinská historical school, and Slovak Mining Museum in Banská Štiavnica. Very close cooperation was established with CA Terra Montanae, which covers the Barbora route project. Subsequently, in August 2023, a Memorandum of

Cooperation was signed with the representatives of CA Terra Montanae, not only in the creation of an interactive platform.

There was excellent collaboration in the development of the interactive platform with CA Štiavnický Tajch. This organisation, along with its dedicated volunteers, is actively involved in the revitalisation of the Banská Štiavnica water management system, which stands as the most significant water management structure ever constructed in Slovakia.

In total, the creation of the "Mining Treasures of Central Slovakia" interactive platform involved more than 40 organisations spanning the public, private, and non-profit sectors. Additionally, 90 students from five academic courses, 27 photographers (including four professionals), and 22 active content authors contributed to the platform. Throughout the development process, over 45 meetings were held with these contributors.

### 8.3.3    *Creation and design of responsive platform—technical part*

The development of the platform is carried out in three main phases: information architecture/web design/development; integration/framework/SEO/training; back-end configuration/implementation.

Based on the specified requirements, during phase 1 (information architecture/web design/development), the information architecture of the interactive platform was developed, followed by the design of the necessary subpages, modules, and elements, and the front-end and back-end was deployed.

Phase 2 included integration of a new visual identity and design modifications; deploying demo content—activities; articles, static pages SEO; preparation of the content framework for preparation; document with manual; cloud content architecture, two-phase training in working with the content framework and deploying content via TYPO 3.

During phase 3 (back-end configuration/implementation) was realised configuration of content management system for language mutations; configuration of recording elements for translation (activities, articles, tags, etc.); route interconnection configuration; Google analytics configuration testing; creating language files with a list of static texts/expressions for translation; implementation of texts from language files; additional filters (route complexity, or other parameter).

After the TYPO3 system was configured, it was possible to start creating the first activities—inserting texts, photos, information and running the platform in a test demo version, which was still being modified. Subsequently, the final user version of the platform was launched in December 2022. The Mining Treasures of Central Slovakia platform can be accessed at www.banickepoklady.eu. Detailed instructions for using the Mining Treasures of Central Slovakia platform are available in Borseková et al. (2024) or directly on the platform in the part "Articles"—"Manual for working with the platform Mining Treasures of Central Slovakia."

*Figure 8.3* Final form of the logo of the platform "Mining Treasures of Central Slovakia."

### 8.3.4 *Creation and design of responsive platform—Design and marketing part—logo competition, participatory approach to logo design and creation*

Students from the courses "Basics of Marketing" and "Marketing of Public and Non-profit Sector" actively contributed to the design of the interactive platform. They were given a specific assignment related to the platform's visual identity. The creation of the visual identity for the "Mining Treasures of Central Slovakia" platform was participatory in nature, facilitated through a logo competition. In total, 31 students participated, resulting in 19 proposed designs for the platform's visual identity, logo, and design manual. The competition yielded numerous compelling logo proposals, making the selection of the final logo to represent the "Mining Treasures of Central Slovakia" a challenging task.

The winning logo design consists of the abstract symbolism of mining, a hammer, which is interspersed with a cross as a sign used in the context of marking a place, a goal, etc. (treasure). The logotype refers to the discovery and wandering of "mining treasures." Logo designs, including the winning logo, can be found in Borseková et al. (2024).

The winning logo design was modified by the designers in the final form (Figure 8.3), which is used on the platform, social media, in presentations, materials, and documents. The winning logo was embedded with a pin symbol, which we use to indicate activity on the platform map. The pin sign also appeared in other student logo designs.

### 8.3.5 *Creation of content for the responsive platform—Participatory approach to platform content*

Creating the content of an interactive platform is a rather demanding process, which includes preparation of text obtained from several sources (books, articles, websites, meetings with the community), photography and editing of photos, obtaining information about entrance, opening hours, time required, track length, restrictions. Furthermore, it is necessary to obtain the exact location using Google Maps and the coordinates, which must be verified directly in the field. The content of the interactive platform is created in a participatory manner with the participation of students, representatives of partner organisations, and scientists involved in the project. Thus, a total of 18 students of the Creativity and Culture in Regional Development course were involved in the creation of the content of the interactive platform, who had the task of visiting selected activities of the "Mining Treasures of Central Slovakia" (museums and galleries, material monuments, industrial monuments) and processing the content of the interactive platform.

Twenty-seven authors and 31 photographers took part in the participatory creation of the content of the interactive platform "Mining Treasures of Central Slovakia."

More than 40 other students of the Creativity and Culture in Regional Development course participate in the promoting of the platform by publishing posts on the social networks Facebook and Instagram using the hashtag #banickepoklady. These are mainly photos from visiting platform activities (usually galleries and museums, monuments, events, etc.)

### 8.3.6 Outputs of participatory creation of the interactive platform "Mining Treasures of Central Slovakia"

A total of more than 100 participants were involved in the participatory creation of the platform "Mining Treasures of Central Slovakia," who took part in the creation and design of the responsive platform within the technical part, design and marketing part, and content development.

During one year of participatory creation of the platform "Mining Treasures of Central Slovakia," more than 45 meetings were organised with more than 40 organisations and communities involved in the creation of the platform. A total of 250 activities were created by 27 authors on the platform. Twenty-seven photographers created more than 2,300 photos for the platform. A total of 90 students in five study subjects were involved in the creation of the content of the platform. Eighteen students prepared content and photos for 54 activities on the platform. Thirty-one students worked on creating the logo, creating a total of 19 logo designs. Forty-one students are still working on promoting the platform through social networks. Currently, the interactive platform has more than 1300 users from 96 countries of the world and their number is constantly growing. Table 8.2 shows the results of the participatory creation of the interactive platform "Mining Treasures of Central Slovakia."

The "Mining Treasures of Central Slovakia" platform successfully ensures the accuracy and authenticity of its content through a series of meticulously designed mechanisms. Collaborative creation is at the forefront, with over 40 organisations and communities actively participating in 45 meetings, contributing diverse perspectives and expertise to the content. The central role of a dedicated coordinator is crucial in this process, overseeing the verification of all submitted content, maintaining communication with stakeholders, and ensuring adherence to cultural and factual standards. Additionally, the platform provides comprehensive training and clear guidelines to all contributors, which helps in maintaining a consistent quality of content. Furthermore, the integration of advanced technology, such as digital archives and GIS mapping, plays a significant role in verifying geographical and historical

*Table 8.2* Results of participatory creation of the interactive platform "Mining Treasures of Central Slovakia"

| Indicator | Value |
| --- | --- |
| Number of organisations and communities involved in the platform creation | 40+ |
| Number of meetings with partners | 45 |
| Total number of students involved in the platform creation | 90 |
| Number of students involved in creating the content of the Mining Treasures website | 18 |
| Number of activities processed by students on the Mining Treasures platform | 54 |
| Number of students involved in the creation of the Mining Treasures logo | 31 |
| Number of Mining treasures logo designs | 19 |
| Number of study subjects in which students participated in the platform creation | 5 |
| Number of activities on the platform Mining Treasures | 250 |
| Number of photos created for the platform | 2300 |
| Number of photographers involved in the creation of web content | 31 |
| Number of authors involved in the creation of web content | 27 |
| Number of social media platforms involved (Facebook, Instagram, and YouTube) | 3 |
| Number of users of the platform | 1300+ |

*Source:* Own elaboration.

data, thereby enhancing the platform's reliability and educational value. These combined efforts result in a platform that is not only rich in diverse inputs but also anchored in accuracy and authenticity, reflecting the true essence of Central Slovakia's cultural heritage.

In addition to its focus on accuracy and authenticity, the "Mining Treasures of Central Slovakia" platform also implements strategies to minimise environmental impacts, promote responsible tourism practices, and enhance community well-being. A key initiative in this regard is the incorporation of educational content about environmentally significant landmarks, such as the Banská Štiavnica Water Management System. This content not only educates tourists about the region's unique environmental features but also highlights ongoing participatory renewal efforts. By informing visitors about the importance of these ecosystems and the role they play in local heritage, the platform encourages more responsible and environmentally conscious tourism behaviours. This approach not only aids in the preservation of these natural systems but also fosters a deeper connection between tourists and the local community, contributing to the overall well-being and sustainability of the region.

## 8.4   Conclusion

Central Slovakia's historic mining locations, connected by the Barbora and Fugger routes, hold immense potential for fostering a heritage community united by a shared cultural legacy. This community should be anchored in the mining history of the region, with a focus on preserving its rich cultural heritage and supporting enduring cultural traditions. The act of preserving and promoting this cultural heritage not only maintains the region's cultural diversity but also reinforces a sense of belonging and identity among its inhabitants. It evokes pride and offers a platform to showcase these values to the broader world.

By following the path of sustainable cultural tourism in Central Slovakia, particularly its historic mining locations linked by the Barbora and Fugger routes, offer a unique opportunity to cultivate a heritage community rooted in the region's rich mining history. This approach emphasises the preservation and promotion of the region's cultural heritage and enduring traditions, thereby maintaining its cultural diversity. By doing so, it not only reinforces a sense of belonging and identity among the local inhabitants but also instils a sense of pride in their cultural legacy. Such sustainable tourism initiatives provide a platform to share and showcase the region's unique cultural values and traditions with a wider audience, balancing the enhancement of regional development and tourism with the conservation of cultural heritage and local community engagement. This aligns with the broader concept of sustainable cultural tourism, which aims to integrate the management of cultural heritage and tourism in a way that benefits all stakeholders and supports both tangible and intangible cultural heritage.

However, despite the region's vast potential and the presence of numerous cultural institutions, such as galleries and museums, as well as civic associations and mining groups, there remains a gap. These entities, all bound by the rich tapestry of mining history, have yet to fully unite and form a cohesive, resilient heritage community. Such a community would not only be a guardian of our ancestors' cultural heritage but would also harness it for territorial development, creating innovative products for sustainable tourism. Central to building this heritage community is the adoption of a participatory approach, one that is rooted in a "bottom-up" perspective.

The "Mining Treasures of Central Slovakia" platform stands as a testament to the power of collaborative effort and the potential of participatory approaches in fostering sustainable cultural tourism. Here's a breakdown of the platform's creation and its impact:

- **Collaborative Creation**: Over 40 organisations and communities came together, holding 45 meetings to shape the platform. This extensive collaboration underscores the importance of collective effort in building something of value.
- **Student Participation**: A significant 90 students from five different academic courses contributed to the platform's creation. Notably, 18 students were directly involved in curating the content for the website, processing 54 activities. Additionally, 31 students participated in a competition to design the platform's logo, resulting in 19 unique designs. This

active involvement of the younger generation highlights the platform's commitment to inclusivity and the blending of academic learning with practical application.

- **Rich Content**: The platform boasts 250 activities, enriched by a staggering 2,300 photos. This vast repository of content was made possible by the contributions of 31 photographers and 27 authors, emphasising the platform's comprehensive approach to showcasing Central Slovakia's mining heritage.
- **Digital Outreach**: The platform's digital presence spans three major social media platforms: Facebook, Instagram, and YouTube. This wide digital footprint ensures that the rich history and cultural significance of Central Slovakia's mining heritage reach a broad audience.
- **User Engagement**: Since its inception, the platform has attracted more than 1300 users, indicating its growing popularity and the potential to become a significant tool for promoting sustainable cultural tourism.

In conclusion, the "Mining Treasures of Central Slovakia" platform exemplifies how a participatory approach can harness diverse talents and resources to create a comprehensive digital tool. Promoting the rich mining heritage of Central Slovakia not only educates and informs but also paves the way for sustainable cultural tourism in the region.

Moving forward, the journey of the "Mining Treasures of Central Slovakia" platform is far from over. The "Mining Treasures of Central Slovakia" is embarking on an exciting phase of development, bolstered by the pursuit of a matching grant from the Slovak Renewal Plan. This project's core objective is to leverage cultural heritage as a key driver for social and economic development in the region, using innovative and original methods that emphasise participation, education, and sustainable tourism. One of the key initiatives in this new phase is the development of an original prototype of a physical map of Central Slovakia and the Barbora Route. This map is unique in that it features a layer designed to connect with an app that integrates the content from the banickepoklady.eu platform. This integration represents a novel approach to tourism, blending traditional and digital methods of exploration. The physical map, paired with the digital app, will allow users to discover traditional sites while simultaneously engaging with digital content, enhancing the overall experience of exploring Central Slovakia's cultural heritage. For this, project is crucial to ensure that the benefits of the platform are distributed fairly among all stakeholders. Furthermore, the expansion of the participatory platform to include Kremnica and its surroundings marks a significant step in the project's development. By employing a participatory approach, similar to what was used for Banská Bystrica and Banská Štiavnica, the platform ensures that the development of new content is inclusive and reflective of the community's needs and insights. This expansion is not only about adding new locations but also about strengthening cooperation with local stakeholders, particularly in Kremnica, and utilising the content for educational purposes. Utilising the content of the platform for educational purposes necessitates the involvement and cooperation of other stakeholders, including educational institutions and their teachers, pupils, students, civic associations, and cultural institutions. They will participate in both the preparation of the educational content for the platform and serve as recipients of its content.

The participatory approach, with its inherent dynamism, serves as an inexhaustible wellspring of inspiration. It continually brings forth fresh ideas and novel avenues for showcasing and leveraging the rich cultural heritage of Central Slovakia. As we continue to embrace this approach, we are not just limited to Central Slovakia but have the potential to set a precedent for regions beyond, illustrating the transformative power of collective effort in celebrating and preserving cultural legacies.

In essence, participatory approaches in tourism serve as a vital instrument for realising sustainable tourism in Central Slovakia. They harmonise economic, cultural, and environmental objectives with the desires and needs of local communities, ensuring the enduring success of tourism destinations. This approach is in line with the principles of sustainable cultural tourism, as it seeks to balance the promotion and preservation of cultural heritage with the economic and social development of the region. The emphasis on participatory methods ensures that the local community is not only a beneficiary but also an active participant in the development process, thus ensuring the sustainability and authenticity of the initiatives. This fact is also a basic prerequisite for the longevity and sustainability of the Mining Treasures of Central Slovakia platform. Long-lasting accessibility of the platform will be ensured by the Local Destination Management Organisation Central Slovakia, with which very close cooperation is developed, and which also actively participates in the creation of the platform's content. Additionally, considering that the Local Destination Management Organisation Central Slovakia is within the organisational scope of the Banská Bystrica self-governing region, sufficient financial and organisational support will be ensured for the long life of the platform.

## Acknowledgements

This chapter has been developed within the framework of the INCULTUM project, funded by the European Union's Horizon 2020 research and innovation programme under Grant Agreement No. 101004552.

## References

Borseková, K., Bitušíková, A., & Vitálišová, K. (2022). *D4.1 Report on participatory models*. Available at: https://incultum.eu/wp-content/uploads/2023/01/INCULTUM-D4.1-new.pdf

Borseková, K., Bitusikova, A., & Rojíková, D. (2024). *INCULTUM mining treasures of Central Slovakia innovation factsheet*. Zenodo. https://doi.org/10.5281/zenodo.10925981

Borseková, K., Vaňová, A.& Petríková, K. (2015). From tourism space to a unique tourism place through a conceptual approach to building a competitive advantage. In A. Correira, J. Gnoth, M. Kozak & A. Fyall (Eds.) *Marketing Places and Spaces (Advances in Culture, Tourism and Hospitality Research, Vol. 10,* pp. 155–172). Leeds: Emerald Group Publishing Limited. https://doi.org/10.1108/S1871-3173201500000010012

Borseková, K., Vaňová, A., & Vitálišová, K. (2017). Smart specialization for smart spatial development: Innovative strategies for building competitive advantages in tourism in Slovakia. *Socio-Economic Planning Sciences*, 58, 39–50. https://doi.org/10.1016/j.seps.2016.10.002

Borseková, K., Vitálišová, K., & Bitušíková, A. (2023). *Participatory governance and models in culture and cultural tourism.* Banská Bystrica, Slovakia: Belianum. https://doi.org/10.24040/2023.9788055720838 Callot, P. (2013). Slow tourism. In S. O. Idowu, N. Capaldi, L. Zu, & A. D. Gupta (Eds.), *Encyclopedia of corporate social responsibility*, pp. 2156–2160. Berlin, Heidelberg: Springer.

European Commission. (2019). *Directorate-general for education, youth, sport and culture.* Sustainable cultural tourism. Publications Office. Available at: https://data.europa.eu/doi/10.2766/400886

European Commission. (2022). *Shaping Europe´s digital future.* Available at: https://digital-strategy.ec.europa.eu/en/policies/cultural-heritage

# 9 Water heritage and community-based cultural tourism. The case of the Algarve's Coastal Agrarian plain, in Southern Portugal[1]

*Desidério Batista, Manuela Guerreiro,
Miguel Reimão Costa, Bernardete Sequeira,
Marisa Cesário and Dora Agapito*

## 9.1    Introduction

This chapter focuses on the role and importance of water heritage in the development of community-based cultural tourism. The study area corresponds to the agrarian plain between the historic cities of Faro, Loulé and Olhão on the Central Algarve Coast. The historical process of building and transforming the landscape reveals the Mediterranean heritage and the Arab-Muslim influence both on the spatial organisation and management models and on land exploitation. These models historically interlink with the interdependence between cities and their food production hinterlands. Vegetable and fruit production depended above all on the efficient usage and management of groundwater for irrigating vegetable gardens and orchards.

Hence, this coastal plain is dominated by irrigated fields with the traditional irrigation system including a wide variety of hydraulic structures. These structures incorporate the lifting of water (*norias*), its transportation (aqueducts), storage (tanks) and distribution (canals) to constitute the basic irrigation system of this ancient and rich agricultural region. Although this water heritage has long since experienced abandonment and degradation, and with traditional cultivation and irrigation practices having fallen into disuse, they nevertheless remain of great historical, cultural, social and environmental interest and value. Therefore, the core challenge of this study is to inventory, rehabilitate and enhance the value of this tangible and intangible heritage while furthermore integrating it into cultural routes that endow visibility and prominence. This agrarian plain reflects a historical area of food production on the fringes of the region's recent urban-tourist development. Furthermore, the area still preserves a unique historical irrigation system of great cultural value spanning several hundred *norias*. The selection of this case study aims to consider the conservation of this valuable heritage and to integrate it into

DOI: 10.4324/9781003422952-9

cultural tourism routes within the context of the transformations ongoing in the contemporary landscape.

This chapter correspondingly comprises of two essential but interrelated and interconnected parts. This sets out by identifying and characterising the historical irrigation system, approaching this as an agrosystem with a long tradition with the *noria* playing a prominent role as a central feature in the local socio-economy and cultural landscape. The chapter proceeds by considering how this valuable hydro-agricultural heritage might serve as a basis for community-based cultural tourism. The chapter then details the participatory-collaborative approaches and innovative solutions coupled with a proposal for cultural itineraries, centred on local communities as communities of practice through their involvement in managing both natural and cultural resources and the hydraulic heritage.

The methodology applied in this research adopts an interdisciplinary, integrative and relational approach that brings together themes from History, Geography, Agronomy, Sociology, Economics and Tourism. This incorporated the cross-referencing of bibliographical and documentary sources with analysis of historical cartography and fieldwork to map and survey hydraulic structures in addition to collecting oral information on traditional cultivation and irrigation practices and techniques.

Therefore, the proposed cultural routes arise out of the landscape and its water heritage as interpreted by the community that manages them, thereby reflecting the participatory-collaborative model. This simultaneously strives to integrate sustainable production and consumption into cultural tourism and hence contribute to improving not only the quality of the environment but also the quality of life of the resident and visiting populations.

## 9.2   Identification and characterisation of the historical irrigation system

The *Gharb al-Andalus* region forged a unique agrarian landscape through its mastery of water for irrigating gardens and orchards (Lagardère 2006). The national and international scientific literature points to the Algarve's historical irrigation system as an Arab-Muslim legacy (Caldas 1998; Feio 1949; Lagardère 2006; Mabberley and Placito 1993; Ribeiro 1991; Schiøler 1973; Stanislawski 1963) that deployed regionally significant technological innovations (Batista 2023; Cavaco 1976; Rodrigues and Merlos 2020).

Alongside the bibliographical sources, primary sources and fieldwork proved fundamental to identifying and characterising the hydraulic structures in the study area. On this agrarian coastal plain, as in other Mediterranean regions, the type of irrigation applied to water vegetable gardens and orchards has been irrigation by shortage (Ribeiro 1991). This type of irrigation requires the most complex and diversified means and procedures for obtaining and distributing water (Ribeiro 1991). Here, the device operated to

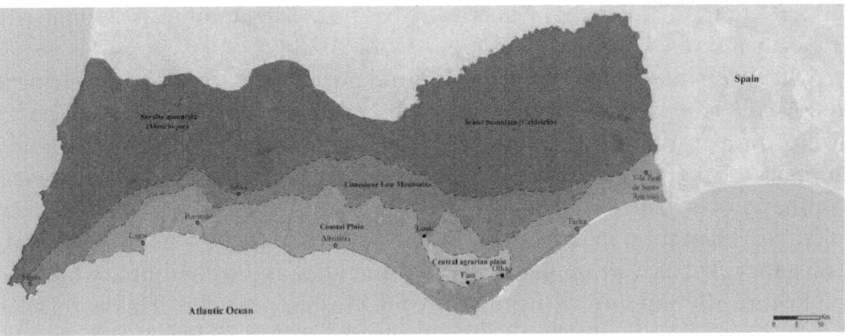


*Figure 9.1* Algarve's map with the location of pilot case. Elaborated by the authors.

lift irrigation water predominates, the *noria* pots, almost always associated with aqueducts, tank(s) and canals.

### 9.2.1 *The Mediterranean legacy: landscapes, practices and techniques*

Written and documentary sources from different periods reiterate the agricultural wealth and extraordinarily productive character of this fertile plain (Cavaco 1976; Feio 1949; Link 1808; Magalhães 1988; Proença 1927; Ribeiro 1991). Historically, it has been a production landscape associated with both rainfed and irrigated agriculture (Batista et al. 2023). Each is associated with a characteristic type of landscape that depends on the fertility of the soils and the presence of water. The former type, associated with less fertile soils, corresponds to the rainfed orchards of almond, fig, carob and olive trees, formerly accompanied by arable crops and legumes (Feio 1949). The latter type of landscape, associated with the greater soil fertility and the availability of water for irrigation, includes vegetable gardens and citrus groves.

One and the other testify to two privileged expressions of the unity and diversity of the landscape of this coastal plain as a legacy of Mediterranean culture (Lagardère 1993; Ribeiro 1991). Its agrarian landscape is particularly marked by the presence of trees in the fields (whether with crops under their cover), considered one of the most characteristic features of the Mediterranean landscape (Ribeiro 1991), alongside vegetables to configure a tight, strongly identifiable and diversified agricultural mosaic.

The Mediterranean polyculture characteristic of this coastal plain combines rainfed and irrigated crops, depending on market fluctuations. This has also experienced the integration of new plants over time as well as developments to the cultivation and irrigation practices and techniques. Rainfed land (traditional mixed orchards, legumes, vines, etc.) and irrigated land (vegetables and fruit) have co-evolved in accordance with a strategy of constant socio-spatial, technological and market adaptation. This process has resulted in the rich and diverse agrarian and hydraulic heritage that fuses Mediterranean heritage with a locally inscribed territorial culture (Batista 2023).

The diversity of fruit and vegetable and nut production that complements the Mediterranean agrarian trilogy of bread, wine and olive oil reflects the legacy of a traditional economy that is simultaneously self-sufficient and a supplier of markets internationally. However, it is also the Mediterranean synthesis of a landscape of mixed cultures that encounters its most complete representation in irrigation (Ribeiro 1991). Here, it is possible to find a constancy of cultivation and irrigation practices and techniques, which can be historically interpreted from the Arab-Muslim presence. This is reflected in the intense development of every form of small-scale agrarian hydraulics based on the mastery of water, resulting in landscapes capable of surviving (Bazzana and De Meulemeester 2009). In the 11th century, Ibn Bassal described the *noria* as a standard machine for irrigation and it remained in use until only recently on the Iberian Peninsula (Al-Hassan and Hill 1986). Indeed, some vegetable gardens on the outskirts of Faro and Loulé still make recourse to the practices of Moorish times with their *norias*, irrigation channels, vegetable plots and orchards (Ribeiro 1991).

The 10th–12th-century chronicles written by Arab historians and geographers (Coelho 1989; Domingues 1971) describe the agrarian landscape of the Algarve coastline and thereby enabling us to identify characters and features that have maintained formal and functional relations through to the present day. Among these descriptions, we would highlight that of Al Râzî (10th century) and Ibn Ghâlib (12th century), which describe the Algarve coastline as flat and fertile, with excellent agriculture, many trees and good seedbeds, very good fruit and watered gardens. The *noria* pots here represent one of the essential features of irrigated agriculture as a medieval Islamic legacy and with little difference in the products grown (Bazzana and Montmessin 2006; Caldas 1998; Lagardère 2006; Mabberley and Placito 1993; Schiøler 1973; Stanislawski 1963).

### 9.2.2   Noria(s) *as an identity expression of water mastery*

The animal-drawn waterwheel (*sâniya* in Arabic, *noria de sangre* in Spanish, *nora* in Portuguese) constitutes one of the main irrigation structures in the Mediterranean world and particularly in the Muslim West (Spain, Portugal, Morocco) (Bazzana and Montmessin 2006). This artefact did not arrive here in isolation but rather as an integral component for a body of technical knowledge, machines and tools linked to new agricultural practices in which irrigated spaces take a central role (Poveda 2004). This "Green Revolution" was based on the *noria*, which allowed farmers to irrigate individual plots and cultivate a wide variety of plants, important both for the economies of medieval towns and for feeding their inhabitants (Glick 1977).

While the Treatise on the Exploration of Groundwater, dated 1017, by the hydraulic engineer Al-Karajî refers to such systems, the perfecting of the animal-drawn bucket gear and its first drawn record are attributed to Al-Jazarî (12th century) (El Faïz 2018). Nevertheless, the spread of *norias* on the Iberian Peninsula is believed to have occurred in the 9th century (Poveda 2004), with the presence of *norias* in the Islamic tradition in the Algarve noted from the 10th century onwards (Lagardère 2006). In whatever the case, the clay pots on display in the Silves Museum date to the Almohad period (12th–13th centuries). The description of how this device for lifting water for irrigation worked by the Sevillian Ibn al-Awwãm, in his Treatise on Agronomy, also belongs to this same period (Caro 1955; Lagardère 1993).

Although the principle associated with its gearing is modest (Glick 1977), they nevertheless represent complex and fragile machines (Bazzana 1994). *Norias* are devices for lifting water in pots (*alqaduzes*, in Arabic, *alcatruzes* in Portuguese) which, interlinked with a well, are moved by a hydraulic wheel powered by the continuous circular movement of animals (donkeys, mules, bulls, etc.). In its most basic version, *norias* in the Algarve featured two wheels, with one horizontal and propelled by the animal and with the second-placed vertically, geared to the first, serving to attach a rope of buckets lifting the water to pour into a canal or tank. This model of animal-drawn water lifting machine is the same as that described in other studies carried out on the Iberian Peninsula (Argemí et al. 1995; Caro 1955; Schiøler 1973; Dias and Galhano 1986; Poveda 2004), and on Islamic technology (Al-Hassan and Hill 1986; El Faïz 2018).

Among the various typologies of *norias* inventoried and characterised in Portugal (Dias and Galhano 1986), this solution was the most common in the Algarve until the end of the 19th century. The vegetable gardens on the Algarve's coastal plain, watered by these *norias*, had hitherto remained small gardens that met the main needs of the local populations for fresh vegetables and fruit (Cavaco 1976).

With the arrival of the railway in the city of Faro in 1889, enabling the subsequent increase in the sale of fruit and vegetable products in other markets, the irrigation processes were both intensified and transformed. This generated implications with an increase in irrigated areas drawing on the heightened utilisation of underground water resources and not only a multiplication of the number of *norias* but also their complexification (Batista 2023; Cavaco 1976). The technological evolution of the waterwheels included replacing the wooden gear by a metal gear, which began then and gathered pace over the first half of the last century. This contributed to the generalisation of a new *noria*'s typology in which the animal's movement was no longer concentric in relation to the wheel but rather in relation to a small structure built alongside. Subsequently, this transformation of the *noria* also extended to integrating reinforced concrete structures in place of pre-existing traditional constructions.

The construction techniques in iron and concrete, and the ability to sink wells further, particularly facilitated the expansion of irrigated areas. This became especially evident following the introduction of a new type of irrigation system, centred on the motor-driven high *noria*. This solution, based on knowledge of agricultural hydraulics, henceforth became widespread and correspondingly associated with irrigated agriculture on an industrial scale and producing an indelible mark on the landscape.

This technological evolution in the waterwheels, incorporating new materials and construction processes, also meant that animals came to be replaced by irrigation motors and electric pumps (Batista 2023; Cavaco 1976). These new methods of lifting water for irrigation gradually dispensed with the feeding of gear and replaced by the opening of artesian boreholes in the last decades of the 20th century. The history of the *noria* tends to progressively close with the advent of this process even though realisation of their importance as a factor of cultural identity and their relative conservation by local communities means that the diversity of this heritage can still today be inventoried and recorded. This is water-based heritage of historical, cultural and environmental values and of particular interest and significance in this context. This stems from not only the large concentration of *norias* pots (with over 400 mapped so far) but also the prevailing typological diversity correspondingly produced by a historical, evolutionary and adaptive irrigation system.

## 9.3    Hydro-agricultural heritage and sustainable cultural tourism

Sustainable tourism planning must consider the four dimensions of sustainable development: economic, social, cultural and environmental sustainability (Bramwell and Lane 2000). Achieving sustainability and the Sustainable Development Goals established by the United Nations requires participatory-collaborative approaches and adopting innovative solutions represents another essential prerequisite. Participation embodies "a process of involving all stakeholders (local government officials, local citizens, architects, developers, businesspeople, and planners) in such a way that decision-making is shared" (Haywood 1988). Thus, a participatory-collaborative approach nurtures a balance between different stakeholders and strives to achieve win-win outcomes in tourism contexts (Arnstein 1969; Ozcevik et al. 2010). Actively involving citizens and local communities in decision-making processes ensures the maintenance of traditional lifestyles and respect for community values (Cater 1994; Murphy 1985; Nunkoo and Ramkissoon 2011; Wild 1994).

This approach derives from a more democratic and inclusive governance model aligned with the emerging smart territory paradigm which perceives the application of technology and the participation of citizens as co-creators in decision-making processes as decisive for achieving both quality of life and quality of the environment (Simonofski et al. 2020).

Within this framework, the participation of the local population in tourism development and heritage management may therefore contribute to improving their own quality of life and, simultaneously, boosting the very sustainability of the territory (Nicholas et al. 2009), strengthening social relations and preserving cultural heritage and the landscape.

Thus, sustainable cultural tourism, understood as heritage-based tourism, supports integrated and sustainable development, especially in neglected and/or peripheral areas (Ottaviani et al. 2023). Similarly, this hydraulic heritage has become a symbol of the identity of territories and with the heritage valorisation of *norias* capable of producing a cultural resource associated with water routes for the diversification of tourism in various contexts (Gil et al. 2020).

Historical irrigation systems are potentially able to provide several different ecosystem services to society (Civantos et al. 2021), which include cultural services related to recreation and leisure (Batista et al. 2023). This reflects the perspective prevailing behind the proposal for cultural routes focused on the hydro-agricultural heritage of the Central Algarve coastal plain in association with sustainable production and consumption practices and participatory tourism.

### 9.3.1  *Participatory-collaborative approaches and innovative solutions*

The participatory-collaborative approach is detailed by international-level documents on sustainable tourism (UNWTO 2005) and academia (Currie et al. 2009; Timothy 2010; Vernon et al. 2005). In the tourism literature, participatory-collaborative approaches enable the sharing of experiences to the benefit of communities (Vernon et al. 2005), which itself assumes an active and decisive role in the planning, development and management of local resources and the respective tourist products (Simpson 2008). This impacts, and among other factors, the development of new products in accordance with UN's sustainable development goals. Furthermore, this approach contributes to promoting capacity-building practices, optimising the multiplier effects of tourism across local communities and increasing general support for the development of local tourism while sharing responsibilities among stakeholders (Bramwell and Lane 2000). The participation of local communities allows tourists to interact with communities and learn about their culture, habits, and the surrounding cultural and natural heritage during their stays (Lucchetti and Font 2013).

In the pilot case on the Algarve's agrarian coastal plain, local communities and stakeholders constitute not only the main actors but also the ultimate beneficiaries and play a direct and important role in the implementation of the pilot actions (Batista et al. 2023). In this pilot, the following participatory-collaborative approaches were adopted: community-based tourism, training and education activities and cultural participation.

### 9.3.2  *Community-based cultural tourism*

In the context of sustainable tourism development, the importance of community-based cultural tourism has been clearly recognised since the 1990s (Salazar 2012). This accounts for a particular and alternative mode of tourism that suggests a mutual relationship in which the tourist does not take central priority but rather represents an equal part of the system (Wearing and McDonald 2002).

Communities and their identities are constantly being reconstructed and redefined due to the way they live and interact, both with each other and with other actors (Waterton and Smith 2010). They are social actors who share cultural and collective identities, historical ties and emotional connections to local heritage and the place where they live (Chitty 2011).

The participatory-collaborative approach therefore aligns with the concept of community-based tourism in which residents, especially in rural communities, are empowered with the ability to manage local tourist resources; generate profit; diversify the local economy; preserve cultural, social and environmental heritage; and generate innovation, among other aspects. Their narratives, through storytelling, inform the design of innovative and authentic tourism experiences (Moscardo 2020). This storytelling makes it possible to collect data based on the stories told (Moscardo 2020) and with participatory Story Maps assisting in recognising the different dimensions of the territory, thereby enabling the georeferencing of the information told and collected.

This participatory-collaborative approach is especially applicable to sustainable heritage tourism contexts where a balance between conservation and tourism needs achieving to reduce and offset negative impacts (Timothy and Boyd 2006). Community-based tourism should benefit from the support and participation of the local community with economic benefits correspondingly distributed among local inhabitants while projects should prioritise protecting the cultural and environmental identities of such locations (Wacław et al. 2015).

This understands the importance of the local community as the main actor and decision-maker in the planning, development and management of the resources necessary to serving the purposes of tourism (Simpson 2008). Within this perspective, we may recognise the key role of the hydro-agricultural heritage of the Algarve's coastal plain, and the participation and cooperation of local farmers and producers as an integral facet of the tourism product and the destination itself. In this sense, the project sought to integrate two complementary approaches to surveying intangible heritage in a prospective dimension, specifically the "study and survey approach" and the "applied approach" (Costa et al. 2022).

The fieldwork therefore played a crucial role by collecting oral information on traditional cultivation and irrigation practices and techniques with interviews carried out with elderly farmers, who are faithful custodians of longstanding knowledge and know-how that is now on the verge of disappearing. The study and diagnosis of hydraulic structures of exceptional historical, cultural and environmental values informed the developing recovery projects. These tasks enable the owners to carry out restoration work on the heritage and then organise visits as part of the cultural itineraries.

Participatory cultural routes activate the local economy, which benefits from the involvement of the farming community as producers and managers of visits to their smallholdings. Through such means, traditional, family-based socio-economies become part of the tourist economy, exploiting tourism niches related to hydro-agricultural heritage (agrotourism, rural tourism, slow tourism). This also strives to ensure that this new tourism segment, an alternative to the sun, beach and golf tourism that predominates in the Algarve, makes an important contribution to combating seasonality. Furthermore, this may provide new and different experiences for visitors, whose immersion in the local rural culture will certainly be remarkable (Batista et al. 2023).

### 9.3.3    *Training and education activities*

In cultural tourism contexts, local communities stand out as fundamental and integral components (Ottaviani et al. 2023). The implementation of training and education actions and activities are decisive for the efficient utilisation of human capital and local resources, especially those with unique value, such as cultural heritage (Scheyvens and Biddulph 2018). The act of interpreting heritage is defined as a means of assistance enabling the broad appreciation of heritage (Tilden 2007; Beck and Cable 1997). Or a creative communication process designed to raise the awareness of both the resident and the visiting populations as regards both appreciating and conserving heritage (UNWTO 2011).

This pilot case included the organisation of various training activities, for example, two participatory workshops on how to build cultural tourism itineraries and products. One session explored ideas and intervention strategies for revaluing water heritage as a basis for defining cultural routes. The other focussed on the cultural, economic and environmental values and social and tourist interests in foodscapes based on traditional irrigation and regenerative agricultural techniques. In both cases, best practices were shared on the implementation of cultural routes designed in collaboration with stakeholders and guided by local producers and small tourist businesses.

This participatory service-learning model for hydro-agricultural heritage as a sustainable cultural and tourist resource was also explored in the education activities carried out. Three scientific and cultural events took place, attended by researchers, heritage managers, university students, local entrepreneurs and citizens. These events made an important contribution to exchanging experiences and knowledge about the role and importance of the landscape and water heritage to sustainable development and community-based cultural tourism.

### 9.3.4    *Cultural participation*

The tourism and culture sectors can mutually nurture each other, generating reciprocal benefits, with cultural participation, coupled with culture and cultural heritage, making a fundamental contribution to the attractiveness of tourist destinations (Guccio et al. 2018). In the methodological development of this pilot project, this dimension reflected in the combination of interviews with "guardians of memory" and storytelling, as well as the perception interviews with to residents and visitors subsequently applied to draft the participatory map.

In accordance with the objective of collecting oral information on traditional cultivation and irrigation practices and techniques in the Campina area and understanding the traditional importance of water in the lives of past and present local communities, we opted for a qualitative methodology. This qualitative methodology allows us to gain an in-depth understanding of the dynamics of social phenomena, considering their complexity and a multiplicity of variables (cf. Carmo 2021). This method serves when the aims involve enabling study participants to share their stories and understanding scenarios or contexts from the perspective of the individuals involved in the phenomenon under analysis (Gonçalves, Marques and Gonçalves 2020). Storytelling was applied as a data collection technique based on the narratives of key informants about their experiences of water in the study area. The essence of storytelling is narrative, i.e., the telling of everyday life situations (Fog, Budtz and Yakaboylu 2005) through the memory of past events (Chronis 2012). Narratives hold the potential to create and interpret places just as getting to know a place also means getting to know its stories (Johnstone 1990). To stimulate participant narratives, we drafted a script featuring key themes, specifically approaching water collection methods, water utilisation and the social life around water, both past and present.

The interviews to the farmers revealed the importance of the *noria* in the complex process of historical construction of the landscape in the study area. These changes were very evident in terms of the waterwheel's construction processes and irrigation techniques. At the same time, they extended to the socio-cultural dimension itself, considering the processes of sharing water and the importance of some of these structures as a place of social gathering in the past and an expression of cultural identity in the present. A summary of the aspects revealed by the interviews can be seen in Table 9.1.

We simultaneously advanced with an online questionnaire for residents and visitors in the Central Algarve coastline region, applying a convenience non-probabilistic sample (Batista et al. 2023). The first section of this survey contained a set of questions related to respondent habits as regards using routes as well as their attitudes and perceptions of cultural and water heritage routes with a focus on the *Campina de Faro*. The second section was designed to collect sociodemographic information. The questionnaire was subject to a pretest.

The sample comprises 134 valid questionnaires with most respondent's resident in Portugal (88%), women (around 70%) and aged under 45 (56%). The results demonstrate how almost 70% of respondents frequently visit or walk along trails/routes with this activity generally performed with family or friends. Most respondents (over 85%) agree they like to go hiking/visit

*Table 9.1* Summary of the aspects revealed by the interviews. Elaborated by the authors

| | Past | | | | | | Present | | |
|---|---|---|---|---|---|---|---|---|---|
| Category | Sub-categories | | | FA | FR (%) | Category | Sub-categories | FA | FR (%) |
| Construction and maintenance of water collection structures | Locating water sources | Sticks (*Varas*) | | 4 | 36,4 | Construction and maintenance of water collection structures | Materials | 1 | 20,0 |
| | | To dig - water borehole | | 7 | 63,6 | | | | |
| | Construction processes | Instruments | *Sarilho* | 9 | 81,8 | | | | |
| | | | Hoe - pickaxe | 2 | 18,2 | | | | |
| | | Materials | Stone masonry | 9 | 69,2 | Water collection methods | Water borehole | 16 | 88,9 |
| | | | Concrete | 3 | 23,1 | | Electrical system | 2 | 11,1 |
| | | | Stone masonry and concrete | 1 | 7,7 | | | | |
| | | Construction/maintenance process | | 34 | 100 | Water distribution methods | Drip irrigation | 5 | 83,3 |
| Water collection methods | Technology-driven devices | | | 20 | 46,5 | | Sprinkler irrigation | 1 | 16,7 |
| | Animal-powered devices | | | 14 | 32.6 | | | | |
| | Water reservoirs | | | 8 | 18.6 | | | | |
| | Springs and fountains | | | 1 | 2,3 | Water use | Agriculture | 8 | 61,5 |
| Water distribution methods | Rega por rojo | | | 21 | 43,8 | | Domestic use | 5 | 38,5 |
| | Levada | | | 15 | 31.3 | | | | |
| | Sequeiro | | | 7 | 14.6 | Water-related social life | Water management | 1 | 50.0 |
| | Drip irrigation | | | 4 | 8,3 | | Socialising and festivities | 1 | 50.0 |
| | Sprinkler irrigation | | | 1 | 2,1 | | | | |
| Water use | Agriculture | | | 39 | 68,4 | | | | |
| | Domestic use | | | 18 | 31.6 | | | | |

(*Continued*)

*Table 9.1* (Continued)

| Past | | | | Present | | | |
|---|---|---|---|---|---|---|---|
| *Category* | *Sub-categories* | *FA* | *FR (%)* | *Category* | *Sub-categories* | *FA* | *FR (%)* |
| Water-related social life | Water management | 19 | 47,5 | Solutions to Water scarcity | Dams and reservoir | 6 | 42,9 |
| | Socialising and festivities | 15 | 37,5 | | Projects and infrastructures | 4 | 28,6 |
| | Water Conflicts | 6 | 15,0 | | Rainfed agriculture | 3 | 21,4 |
| | | | | | Water mills | 1 | 7,1 |
| Water scarcity | | 39 | 100 | Water scarcity | | 9 | 100 |

the cultural trails/water heritage routes (e.g., waterwheels, aqueducts, tanks, canals) around Faro. The percentage of respondents who would like to be accompanied by a specialised guide from the community slightly decreases (around 70%) and falls still lower when having to pay for the service (around 52% agree). More than 20% of the respondents are acquainted with the region's water and cultural heritage routes with 16% having already visited this region. Furthermore, over 30% of participants had not visited the area but had already heard about it. In overall terms, the respondents perceive that Campina's surroundings are ideal for establishing thematic routes related to water-based cultural heritage.

The results of these surveys lead us to conclude that the study area, despite being in the most touristic region of the country, remains practically unknown and only on the margins of tourist activities. However, the destination gains recognition among those who know about its great cultural interest associated with historic irrigation practices. This legacy, which embodies a strong identity, reflects the exceptional heritage and environmental value. Interviews with the custodians of memory help tell the story of the usage and management of water. Its (re)discovery through the participatory map and cultural routes associated with community-based tourism will generate an important boost for the local economy while reconciling the objectives of sustainable agricultural production and the enjoyment of water heritage.

## 9.4   Proposed cultural routes as a basis for community-based cultural tourism

Cultural ecosystem services correspond to the intangible benefits that people obtain from ecosystems through spiritual enrichment, cognitive development, recreation and aesthetic experiences, spanning activities such as tourism and the spaces and facilities that are at the centre of daily life (Church et al. 2015), such as agrarian fields.

The Algarve's coastal agrarian plain reflects a historical model of occupation and spatial organisation based on the deep relationships between living and producing, between the house and the garden. Here, a territorial culture prevails, based on daily agricultural labour and experiences with a strong connection to the land and the place. As an everyday landscape, this cultural landscape materially reflects the old knowledge associated with traditional cultivation and irrigation practices.

The proposal for cultural routes on this hydro-agricultural heritage answers three main objectives stemming from the UN-SDGs (8, 11, 12 and 15) (Batista et al. 2023). The first objective involves revaluing the natural and heritage resources associated with traditional agriculture and historic irrigation, reactivating and inspiring a sustainable practice community around landscape routes and water heritage. The second objective is to foster an alternative to the established tourism services, building a sustainable tourism product around local heritage and the Mediterranean diet, targeting those tourists and visitors who appreciate local culture. The third objective is to advance with an integrated territorial project that reconciles regenerative agriculture and participatory cultural tourism as a basis for sustainable regional development.

To fulfil these objectives, the proposal incorporates four key strategies (Batista et al. 2023). The first strives to revive the history and memory of the places associated with traditional irrigated agriculture, making a decisive contribution to increasing agro-diversity and biodiversity. The second strategy involves promoting the production and sale of local products, which are traditionally produced and processed according to local methods and recipes and constitute living examples of the Mediterranean diet, thereby providing a clear contribution to enhancing their value as the intangible cultural heritage of humanity. The third strategy seeks to reinforce the position that the sector's economic profitability overlaps with its collective memory and cultural identity, with lasting advantages for the development of sustainable tourism based on the deep interrelationships between the cultural sector (cultural participation, cultural heritage) and the tourism sector. Finally, the fourth strategy relies on participatory approaches and models that are widely accepted as the criteria necessary for sustainable tourism given how they help decision-makers to maintain traditional lifestyles and respect community values (Wild 1994). Furthermore, they also serve the purpose of developing the image and brand of the tourist destination and increasing its competitiveness, producing both better customer services and innovative tools (Wang and Fesenmaier 2007) for promoting culturally and environmentally based tourist activities.

The proposed hydro-agricultural heritage route is based on the existing valuable set of hydraulic structures and helps to preserve the landscape memories and reactivate its identity. The innovative approach suggested also includes open-air markets selling fruit and vegetables in the villages, which

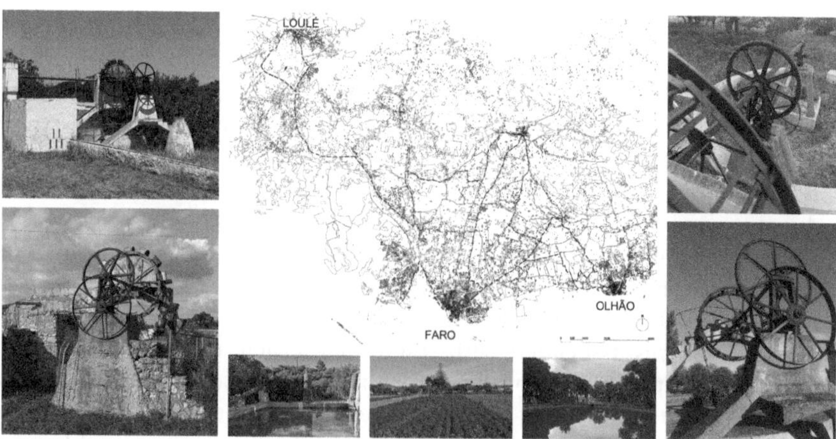


*Figure 9.2* Proposed cultural routes on water heritage based on project's results.

serve as an attraction for visitors, bringing tourism closer to polyculture and the Mediterranean diet and furthermore contributing to the local economy. This also promotes farmer-led tours of the water heritage and foodscapes, providing visitors with new and different experiences within living and authentic communities of practice.

## 9.5    Final remarks

The historical process of building and transforming the landscape of the Algarve's coastal plain reveals ancient and continuous human occupation. The organisation and management model of this agrarian landscape highlights the Mediterranean heritage. Historically, vegetable and fruit productions have drawn on cultivation and irrigation practices and techniques that reveal the Arab-Muslim influence. This is reflected in the mastery of water associated with the efficient utilisation and management of irrigation water based on a complex and sophisticated irrigation system. This system combines tradition, the result of that heritage, which integrates the regional and local innovations and adaptations that took place across different historical periods.

The *noria* buckets, as a fundamental hydraulic structure for fruit and vegetable production, demonstrate the evolution and adaptation of this ingenious irrigation system in the transition from family farming to market farming at the turn of the 19th century. Its historical and cultural interests and values, as well as its importance in terms of the collective identity and landscape the memory, endow the *noria* with a heritage dimension (material and immaterial) that requires safeguarding and valuing in the face of progressive abandonment and degradation. To this end, we propose a network of cultural routes that, by combining production with recreation, enable this landscape and water heritage to be rediscovered and made visible.

This study correspondingly adopted the participatory-collaborative model, which fundamentally includes community-based cultural tourism, training and education activities as well as cultural participation. The innovative approaches on which this approach is dependent comprise primarily of fieldwork undertaken with two main objectives – firstly, the inventorying and surveying of water heritage with a view to its recovery and valorisation, integrated into cultural visits managed by the local community; secondly, the collection of oral information from elderly farmers, the "guardians of memory", whose knowledge and know-how help to tell the story of the landscape and the utilisation of water and thereby contributing to drafting the collaborative maps. These underpin the cultural routes through the hydro-agricultural heritage generating a didactic resource, a tourism product and a vector for territorial and social cohesion.

Studying this heritage involved both the photographic recording of over 200 hydraulic structures and their associated domestic architecture and the surveying of 42 heritage sites and three historical estates in the three municipalities making up the field of study. In accordance with this prior work, the findings produced five cultural heritage routes and two architectural projects for the rehabilitation of the hydraulic structures. The key restraint faced by the project stems from this water heritage being private property. Work recently began on implementing one of the aforementioned routes with the partnership and cooperation of the local community, the local authorities and other stakeholders.

## Note

1  This chapter is part of INCULTUM Project 2021–2024 financed by the H2020 programme of the European Union under Grant Agreement n. 101004552 and is part of the CEAACP project UIDB/00281/2020 funded by FCT (Fundação para a Ciência e Tecnologia, Portugal).

## Bibliography

Al-Hassan, Ahmad and Hill, Donald. *Islamic Technology: An Illustrated History.* Cambridge: Cambridge University Press/Unesco, 1986.

Argemí, Mercé R.; Barceló, Miquel, P.; Cressier, Patrice; Kirchner, Helena G. and Navarro, C. "Glosario de términos hidráulicos.' In López Sánchez, María (Ed.), *El agua en la agricultura de al-Andalus*, 163–189. Barcelona: Lunwerg – El Legado Andalusí, 1995.

Arnstein, Sherry R. "A ladder of citizen participation." *Journal of the American Institute of Planners* 35, no. 4 (1969): 216–224.

Batista, Desidério. "Le paysage et le patrimoine hydraulique comme base du tourisme culturel. Le cas de la Campina de Faro (Algarve, Portugal) dans le cadre du Project européen INCULTUM." *Al-Sabîl: Revue d´Histoire, d´Archéologie et d´Architecture Maghrébines* no. 15 (2023): 1–23.

Batista, Desidério; Guerreiro, Manuela; Sequeira, Bernardete; Cesário, Marisa and Agapito, Dora. "The cultural landscape of Campina de Faro: solutions based on water heritage and cultural tourism." *Revista PH* no. 109 (2023): 88–111.

Bazzana, André. "La pequeña hidráulica agrícola en al-Andalus." *Ciencias de la naturaleza en al-Andalus* III (1994): 317–335.

Bazzana, André and Meulemeester, Johnny De. *La Noria, L´Aubergine et le Fellah. Archéologie des Espaces Irrigués dans L´Occident Musulman Médiéval (9e–15e siècles).* Ghent: Academia Press, 2009.

Bazzana, André and Montmessin, Yves. "Nâ`ura et sânyîa dans l´hidraulique d`al Andalus à la lumière des fouilles de «Les Jovades» (Oliva, Valence)." In Patrice Crescer (Ed.), *La maîtrise de l´eau en al-Andalus. Paysages, pratiques et techniques* (2006): 209–287.

Beck, Larry S. and Cable, Ted. *Interpretation for the 21st Century: Fifteen Guiding Principles for Interpreting Nature and Culture.* Champaign, IL: Sagamore Publishing, 1997.

Bramwell, Bill and Lane, Bernard. *Tourism Collaboration and Partnerships: Politics, Practice, and Sustainability.* Clevedon: Channel View Publications, 2000.

Bryman, Alan. *Social Research Methods*, 5th ed. Oxford: Oxford University Press, 2000.

Caldas, Eugénio C. *A Agricultura na História de Portugal.* Lisboa: Empresa de Publicações Nacionais, 1998.

Carmo, Hermano. "O valor da Investigação Qualitativa na Produção Científica." In Gonçalves, Sónia; Gonçalves, Joaquim and Marques, Célio Gonçalo (Coord.), *Manual de Investigação Qualitativa. Conceção, análise e interpretações*, 17–21. Lisboa: Pactor, 2020.

Caro, Júlio B. "Sobre la historia de la noria de tiro". *Revista de dialectología y tradiciones populares*, no. 11 (1955): 15–79.

Cater, Erlet. "Ecotourism in the third world: Problems and prospects for sustainability." *Ecotourism: A Sustainable Option?* 2, no. 2 (1994): 69–86.

Cavaco, Carminda. *O Algarve Oriental. As vilas, o campo e o mar*. Faro: Gabinete de Planeamento da Região do Algarve, 1976.

Chitty, Gill. "Lived cultural heritage: Southport and beyond, the historic environment." *Policy & Practice* 2, no. 2 (2011): 160–164.

Chronis, Athinodoros. "Between place and story: Gettysburg as tourism imaginary." *Annals of Tourism Research* 39, no. 4 (2012): 1797–1816.

Church, Andrew; Mitchell, Richard; Ravenscroft, Neil and Stapleton, Lee M. "Growing your own`: A multi-level modelling approach to understanding personal food growing trend and motivation in Europe." *Ecological Economics* 110 (2015): 71–80.

Civantos, José M.; Bonet, Teresa and Abellán, José. *Améliorer la cohésion sociale des territoires Méditerranéens autor de l'eau*. Grenade: Université de Grenade, 2021.

Coelho, António B. *Portugal na Espanha Árabe, vol. 1 Geografia e Cultura*. Lisboa: Editorial Caminho, 1989.

Costa, Miguel R.; Gómez, Susana M.; Delgado, Aniceto M.; Costa, Catarina A.; Rosado, Ana C. and Espino, Blanca H. "Sustainable development in rural and peripheral areas through the safeguarding of their immaterial cultural heritage." *Architecture, City and Environment* 50, no. 17 (2022): 1–22.

Currie, Russell R.; Seaton, Sheilagh and Wesley, Franz. "Determining stakeholders for feasibility analysis." *Annals of Tourism Research* 36, no. 1 (2009): 41–63.

Dias, Jorge and Galhano, Fernando. *Aparelhos para elevar água de rega. Contribuição para o estudo do regadio em Portugal*. Lisboa: Etnográfica Press, 1986.

Domingues, José G. "Ossónoba na Época Árabe." In Pinheiro e Rosa and José António (Eds.), *Anais do Município de Faro*, 180–229. Faro: Município de Faro, 1971.

El Faïz, Mohammed. *Os Mestres da Água: História da Hidráulica Árabe*. Faro: Universidade do Algarve, 2018.

Fog, Klaus; Budtz, Christian and Yakaboylu, Baris. *Storytelling: Branding in Practice*. Berlin: Springer, 2005.

Feio, Mariano. *Le Bas Alentejo et l'Algarve*. Lisbonne: Université de Lisbonne, 1949.

Gil, Encarnación M.; Bravo, José M.; Bernabé-Crespo, Miguel and Gómez, José E. "Waterwheels as tourist resource and signs of identity of the Ricote Valley (Region of Murcia-Spain)." *Gran Tour: Revista de Investigaciones Turísticas* no. 22 (2020): 71–96.

Glick, Thomas. "Noria pots in Spain." *Technology and Culture* 18, no. 4 (1977): 644–650.

Gonçalves, Sónia P., Marques, Célio G. and Gonçalves, Joaquim P. *Manual de Investigação Qualitativa. Conceção, análise e interpretações*. Lisboa: Pactor, 2020.

Guccio, Calogero; Mazza, Isidoro; Mignosa, Anna and Rizzo, Ilde. "A round trip on decentralization in the tourism sector." *Annals of Tourism Research* 72 (2018): 140–155.

Haywood, Michael K. "Responsible and responsive tourism planning in the community." *International Journal of Tourism Management* 9 (1988): 105–118.

Johnstone, Barbara. *Stories, Community, and Place. Narratives from Middle America.* Bloomington: Indiana University Press, 1990.

Lagardère, Vincent. *Campagnes et Paysans d'Al-Andalus (VIIIe–XVe S).* Paris: Editions Maisonneuve et Larose, 1993.

Lagardère, Vincent. "Appropriation des terres, maîtrise des eaux et paysages agraires dans le district (iqlim) de Silves (Xe–XIIIe siècles)." In *La maîtrise de l'eau en al-Andalus. Paysages, pratiques et techniques,* 75–111. Madrid: Casa de Velásquez, 2006.

Link, Heinrich F. *Voyage en Portugal, fait depuis 1797 jusqu'en 1799.* Paris: Dentu, Imprimeur-Libraire, 1808.

Lucchetti, Veronica G. and Font, Xavier. "Community based tourism: Critical success factors." *ICRT Occasional Paper* 27 (2013): 1–20.

Mabberley, David J. and Placito, Peter J. *Algarve Plants and Landscape. Passing Tradition and Ecological Change.* Oxford: Oxford University Press, 1993.

Magalhães, Joaquim R. *Algarve Económico 1600–1773.* Lisboa: Editorial Estampa, 1988.

Moscardo, Gianna. "Stories and design in tourism." *Annals of Tourism Research* 83 (2020): 102950–102962.

Murphy, Peter. *Tourism: A Community Approach.* New York: Methuen, 1985.

Nicholas, Lorraine; Thapa, Brijesh and Ko, Yong J. "Resident's perspectives of a World Heritage Site. The Pitons Management Area, St. Lucia." *Annuals of Tourism Research* 36, no. 3 (2009): 390–412.

Nunkoo, Robin and Ramkissoon, Haywantee. "Developing a community support model for tourism." *Annals of Tourism Research* 38, no. 3 (2011): 964–988.

Ottaviani, Dorotea; Demiröz, Merve; Szemzö and De Luca, Claudia. "Adapting methods and tools for participatory heritage-based tourism planning to embrace the four pillars of sustainability." *Sustainability* 15, no. 6 (2023): 1–26.

Ozcevik, Ozlem; Beygo, Cem and Akcakaya, Imge. "Building capacity through collaborative local action: Case of Matra REGIMA within Zeytinburnu regeneration scheme." *Journal of Urban Planning and Development* 136, no. 2 (2010): 169–175.

Poveda, Angel S. "Un estudio sobre las norias de sangre de origen andalusí: el caso de la alquería de Benassal (Castellón)." *Historia Agraria* no. 32 (2004): 35–54.

Proença, Raul. *Guia de Portugal. Estremadura, Alentejo, Algarve.* Lisboa: Fundação Calouste Gulbenkian, [1927] 1991.

Ribeiro, Orlando. *Portugal o Mediterrâneo e o Atlântico.* Lisboa: Livraria Sá da Costa Editora, [1945] 1991.

Rodrigues, Ana D. and Merlos, Magdalena R. "Noras, Norias, and technology-of-use." In Rodrigues, Ana D. and Merlos, Magdalena R. (Eds.), *The History of Water Management in the Iberian Peninsula between the 16th and 19th Centuries,* 331–350. Cham: Springer, 2020.

Salazar, Noel B. "Community-based cultural tourism: Issues, threats, and opportunities." *Journal of Sustainable Tourism* 20, no. 1 (2012): 9–22.

Scheyvens, Regina and Biddulph, Robin. "Inclusive tourism development." *Tourism Geographies* 20, no. 1 (2018): 589–609.

Schiøler, Thorkild. *Roman and Islamic Water-Lifting Wheels.* Odense: Odense University Press, 1973.

Simonofski, Anthony; Van der Storme, Stefanie and Meers, Hanna. "Towards a holistic evaluation of citizen participation in smart cities." In *dg. o'20: the 21st Annual International Conference on Digital Government Research,* 2020, 1–8.

Simpson, Murray C. "Community benefit tourism initiatives - A conceptual oxymoron?" *Tourism Management* 29, no. 1 (2008): 1–18.

Stanislawski, Dan. *Portugal's Other Kingdom.* The Algarve: University of Texas Press, 1963.

Tilden, Freeman. *Interpreting Our Heritage,* 4th ed. Chapel Hill: The University of North Carolina Press, 2007.

Timothy, Dallen J. "Cooperative tourism planning in a developing destination." *Journal of Sustainable Tourism* 6, no. 1 (2010): 52–68.

Timothy, Dallen J. and Boyd, Stephen W. "Heritage tourism in the 21st century: Valued traditions and new perspectives." *Journal of Heritage Tourism* 1, no. 1 (2006): 1–16.

UNEP and UNTWO. *Making Tourism More Sustainable: A Guide for Policy Makers.* Paris: UNEP/Madrid: World Tourism Organization, 2005.

UNWTO. *UNWTO Tourism Highlights, 2004 Edition.* Madrid: World Tourism Organization, 2005.

UNWTO. *Communicating Heritage – A Handbook for the Tourism Sector.* Madrid: World Tourism Organization, 2011.

Vernon, Jon; Essex, Stephen; Pinder, David and Curry, Kaja. "Collaborative policymaking: Local sustainable projects." *Annals of Tourism Research* 32, no. 2 (2005): 325–345.

Wacław, Idziak; Janusz, Majewski and Piotr, Zmyślony. "Community participation in sustainable rural tourism experience creation: A long-term appraisal and lessons from a thematic villages project in Poland." *Journal of Sustainable Tourism* 23, no. 8–9 (2015): 1341–1362.

Wang, Youcheng and Fesenmaier, Daniel R. (2007) "Collaborative destination marketing: A case study of Elkhart County, Indiana." *Tourism Management* 28 (2007): 863–875.

Waterton, Emma and Smith, Laurajane. "The recognition and misrecognition of community heritage." *International Journal of Heritage Studies* 16, no. 1–2 (2010): 4–15.

Wearing, Stephen L. and McDonald, Matthew. "The development of community-based tourism: Re-thinking the relationship between tour operators and development agents as intermediaries in rural and isolated area communities." *Journal of Sustainable Tourism* 10, no. 3 (2002): 191–206.

Wild, C. "Issues in ecotourism." *Progress in Tourism, Recreation, and Hospitality Management* 6 (1994): 12–21.

# 10 Historical water management systems and sustainable tourism

## The case study of Altiplano de Granada (Spain)

*José María Martín Civantos, Elena Correa Jiménez and María Teresa Bonet García*

## 10.1    Introduction

The proposal to create cultural routes using the historic *acequias* (irrigation ditches) as content and vessel arises from work and research with the local communities in charge of their management and maintenance. It is part of the fruit of inter- and transdisciplinary collaborations that have generated a great deal of information that must be transmitted to the population; in this case, through tourist routes. It also arises from the need to value the work of farmers and irrigators in rural and marginal areas. The approach is based on the involvement of local actors in the design and content of these trails, and always for the benefit of the populations and socio-ecosystems, not to their detriment.

These trails, which in fact already exist, as they are the historical paths parallel to the canals along which the acequieros,[1] offer the visitor a wealth of information from different points of view:

A From an archaeological point of view, the "acequieros' paths" are paths full of history; they are paths parallel to the main *acequias* that still have a servitude of way. In general, their layout dates back to the medieval Islamic period, at least a thousand years ago. They are the result of a historical co-evolutionary process in which the local ecological knowledge (LEK) of generations of farmers and shepherds is contained. Therefore, we can offer the visitor abundant historical-archaeological information through written sources, historical cartography, landscape studies and analysis of the hydraulic infrastructures linked to these irrigation systems and the associated human settlements themselves.

B From an environmental point of view, they generate a high level of agricultural diversity (crops, pastures, etc.) and biological diversity (species and habitats). A large amount of flora can be seen along the banks of the *acequia*. They increase the production of crops, pastures and other ecosystems, thus contributing to the food sovereignty of the territory.

DOI: 10.4324/9781003422952-10

*Figure 10.1* Example of the Acequiero's Path (acequia in Lugros, a municipality in the Guadix region, Granada).

C From a geographic point of view, these channels provide information about the settlement, territorial organisation and the organisation of urban and agricultural spaces.
D From the Anthropological study, Irrigation Communities manage these systems on the basis of LEK and communal governance institutions based on historical irrigation rights with complex mechanisms for water distribution and conflict resolution.
E From Hydrogeology, as they regulate hydrological cycles, recharging aquifers and springs, they are water seeding and harvesting systems, examples of Nature-based Solutions (Martos et al., 2020) and Integrated Water Management Resources.

## 10.2   Previous research and work: the study and recovery of agricultural landscapes

The background to this proposal arises from the FP7 MEMOLA Project (2014–2017). One of the main objectives of this project was the evaluation of the historical and long-term use of water and soils in the different study areas. In the case of our research (Geographical context|MEMOLA Project: Mediterranean Mountain Landscapes), it involved the contact and direct participation of Irrigation Communities. Since then, traditional communities have been the focus of our research, not only at the historical-archaeological level, but also by providing them with institutional support, alternatives for organisation and mediation, as well as the promotion of volunteer initiatives and citizen participation in the recovery and cleaning of *acequias* (Delgado, 2017).

Through these proposals, we were not only able to document a large number of historic *acequias* networks and all the elements associated with them, but also to get to know and deal with many farmers and irrigators who have kept these systems in operation to the present day, thanks to a sustainable and efficient way of managing the environment and natural resources. We have learned first-hand about the problems that most affect these historical and traditional irrigation systems and, therefore, threaten their survival: the lack of generational replacement, the strong pressures of a voracious market that does not establish fair prices for small producers, the overexploitation of natural resources by intensive crops, or the contradictions of the administration itself which, on the one hand, applauds conservation initiatives while, on the other, finances millionaire projects of irrigation technification modernisation and destruction of these historical irrigation systems, in pursuit of an intensification and industrialisation of agricultural production with strong environmental impacts to name just a few. Faced with these needs, we promoted the creation of the Association of Historical and Traditional Irrigation Communities of Andalusia ACEQUIAS HISTÓRICAS – General Secretariat. Its objectives are to defend the historical rights of the communities; to support the defence of their territories and infrastructures; to promote social recognition; to promote economic development; to act as an spokesperson with the administration.

This situation of pressure and danger of extinction and destruction, not only of the acequias themselves, but also of all the traditional knowledge associated with their management and administration, led us to promote initiatives to recover and maintain these socio-ecosystems and, consequently, the cultural landscapes they generate, as well as to transfer and compile all the traditional knowledge of water management and administration. These activities for the recovery of historic *acequias* have arisen mainly as a tool for social intervention that aims, above all, to provoke, energise and set in motion participatory processes of reflection and debate linked to action. These initiatives, promoted from the university sphere, seek to reactivate communal functioning mechanisms, through the participation of local and foreign volunteers (university students, environmental and cultural volunteers, etc.), the transmission of knowledge and generational change. Since 2014[2] we have been carrying out this type of activity every year with a large number of participants and with encouraging results.

The choice of the Granada Altiplano as a study area for the INCULTUM project was based on previous work on historical irrigation systems. The Altiplano of Granada is one of the areas in the southeast of the peninsula where the risk of desertification is greatest. In recent years, it has been suffering from a serious exploitation of water resources by intensive crops, causing significant and irreversible damage to small farms, which are the ones that maintain the historical and traditional irrigation systems.

This initial work in the Altiplano was done in collaboration with local corporations, specifically the Rural Development Group (GDR) Altiplano de Granada. Together with the GDR, we were able to make contact with a marginal area that is quite unknown and with an overwhelming historical heritage. The information on the hydraulic heritage collected in this region (historical irrigation systems, hydraulic infrastructures – mills, cisterns –, water sources and springs, etc.) was only a first contact. This project was part of a participatory process that arose from the presentation of arguments to the Important Issues of the Third Cycle of Hydrological Planning. We wanted to draw attention to the importance of this heritage and the need to establish measures for its protection, without destroying the traditional values that have kept it alive to this day. To this end, we drew up a "Catalogue-Inventory of hydraulic heritage" in which we set out the guidelines to be followed in subsequent works, given the enormous number of elements that need to be known and documented in order to be protected. Our work was a small contribution which, together with the great work carried out by the GDR Altiplano, resulted in the proposal of an "Alto Guadiana Menor" River Contract, a large-scale participatory process that will lead to the protection of this fundamental resource in the Altiplano region.

The INCULTUM project has allowed us to broaden the strategies for better conservation, protection and knowledge of the historical irrigation systems of the Altiplano by introducing sustainable cultural tourism as an element that can provide tangible solutions.

## 10.3    Desert landscapes and oasis: Altiplano de Granada

The Granada Altiplano is a territory located in the north of the province of Granada, which includes the districts of Baza and Huéscar. In our case, we have also included a sector of the district of Guadix, which is part of the same geographical entity and which, in fact, is also included within the Granada UNESCO Geopark area.

The geographical peculiarities of this territory have contributed to the creation of a unique landscape marked by impressive badlands where human beings, in relation to the environment, have built balances based on a sustainable use of resources, in particular water and soils. This synergy has generated cultural landscapes in which historical irrigation systems play a fundamental role in the creation of authentic oases of great beauty, with several cultural and environmental values.

The construction of these systems dates mainly from the Middle Ages, during the Islamic period. They are still in use, managed by local farming communities, but are threatened by abandonment and agricultural intensification. The Altiplano is, in fact, part of the so-called "empty Spain", affected by a worrying process of depopulation and with one of the lowest per capita

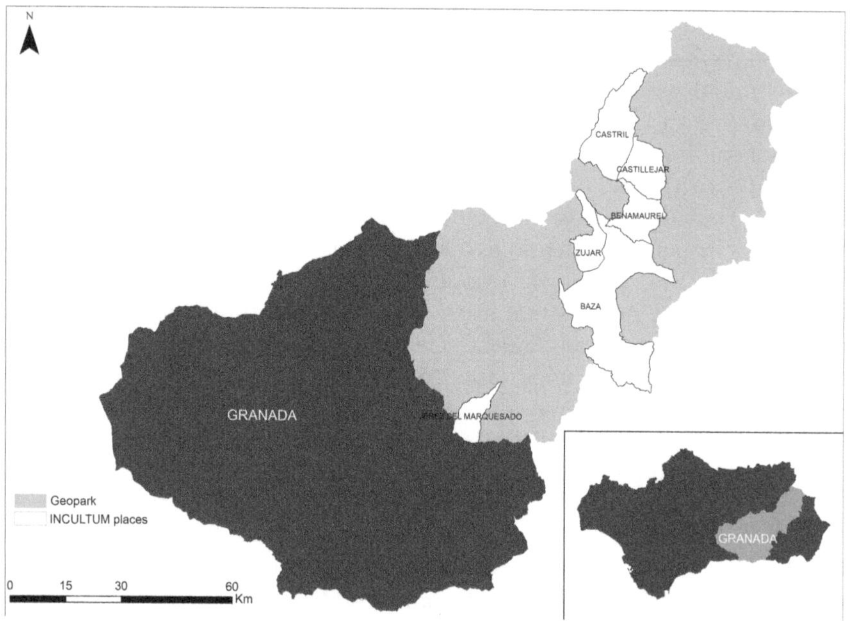



*Figure 10.2* Location of the Geopark territory and the different municipalities in which the INCULTUM proposal has been implemented.

incomes in Spain. However, its enormous cultural and natural potential has led to the recent declaration of this territory as a UNESCO Global Geopark (10 July 2020).

The Granada Geopark covers a total surface area of 4,722 km and comprises 47 municipalities (see map). It has more than 70 sites of geological interest and represents an exceptional place for research from different scientific disciplines and the consequent dissemination and enhancement of all these areas of enormous heritage and geological value.

The figure of "Geopark", in addition to combining a space of great geological, anthropological and historical values, has as one of its main objectives to enhance the value of the heritage of the territory, using it as a resource for the development of the population. It is a way of diversifying resources, helping in many cases to curb depopulation, which is very present in rural and marginal environments, such as the Altiplano area. It is in this exceptional environment where our pilot case is located, although due to its extension, we have focused on the municipalities of Castril, Benamaurel, Zújar, Baza and Jérez del Marquesado.

As in most of the south-east of the peninsula, in all these municipalities, the historical irrigation systems, established in medieval Islamic times, have been shaping the landscapes by taking advantage of the natural resources (water sources, rivers, groundwater, etc.) of each of these places. The semi-desert situation of the Altiplano means that the historical irrigation systems have generated authentic oases that make the most of the water resources to create meadows with a wide variety of crops based on small farms.

Historical and traditional irrigation systems are a fundamental part of our landscapes and our cultural heritage. Their social, territorial, environmental, productive and, of course, cultural significance have been fundamental throughout the last millennium. These systems generate a large number of ecosystem services and have proven to be highly resilient to social, environmental, political and economic changes over time. However, they are not taken into account as unique and fundamental elements for the ecological transition and sustainable development of these regions. Despite their values and the ecosystem services they produce, they are being replaced by pressurised and technified irrigation systems aimed at increasing production for the global market. These new production systems not only tend to rapidly deplete resources, mainly water, through overexploitation of aquifers, but also contribute to the destruction of soils and, with them, of the biodiversity associated with traditional agricultural production.

Historical irrigation systems are managed by farmers' collectives with a wealth of traditional and environmental knowledge, grouped in Irrigation Communities. These are responsible for the management and distribution of water and the maintenance of the infrastructure of these systems. They are extremely complex communal institutions, which face significant economic and demographic problems without recognition of their work by official institutions. The empowerment and dynamisation of these Irrigation Communities is one of the fundamental objectives of our work, as this is the only way they will be able to cope with the administrative obstacles and pressures that they usually have to face.

## 10.4    Social innovation for a sustainable cultural tourism

The actions programmed in the pilot case are carried out by the University of Granada (MEMOLab – Biocultural Archaeology Laboratory) in collaboration with the Provincial Tourism Board of Granada, the Association of Historical and Traditional Irrigation Communities of Andalusia (Acequias Históricas) and the Rural Development Group of the Baza-Huéscar region (GDR Altiplano), as well as town councils and Irrigation Communities and other local associations and groups. The main activity is to convert some of the existing paths adjacent to the main *acequias* into cultural itineraries of great attraction for their landscape, cultural and environmental values, and to link them to local agricultural production, rural heritage, traditional practices and ecosystem services. The aim is to generate a cultural resource based on agricultural heritage and agricultural-based cultural landscapes. It is ultimately based on the multifunctionality of traditional agrarian socio-ecosystems, which generate a significant amount of ecosystem services, including cultural ones. The ultimate objective is to contribute to diversifying the economic activity of irrigation and farming communities and to generate a return directly to them while trying to avoid the negative impacts, not only of touristification in territories and contexts with delicate balances, but also of the intensification and industrialisation of agricultural production.

All this work involves a participatory approach, giving the Irrigation Communities a leading role in the proposal, design and management of the routes, so that they have a direct return, both material and immaterial. Therefore, these routes, which run alongside the historical *acequias,* will also cover part of the cultural landscape, the historical and natural heritage and, of course, the agricultural and gastronomic heritage of each locality.

With this, we intend to make the routes one more service offered by these historical and traditional Irrigation Communities. These corporations have been systematically undervalued and neglected by administrations. Within the pilot, for the correct development and conservation not only of the cultural routes, but also of the irrigation systems and the landscape, we have set up "payment for ecosystem services agreements"; this has been implemented through the signing of administrative agreements between town councils and Irrigation Communities. The agreements have been aimed at facilitating local administrations to fulfil their competences and implement policies that contribute to local development and to the maintenance and improvement of traditional and historical irrigation systems, especially the values and services enjoyed by citizens and the environment. It is also intended to be a tool for the promotion of local tourism, cultural and educational initiatives, fostering new potential sources of employment and economic development, such as, for example, the cultural itineraries associated with *acequias* and irrigated areas. Finally, this agreement aims to institutionalise and give legal coverage to municipal public investments in green and blue infrastructures, providing a reference framework and greater legal security.

The agreements arise as a tool of recognition, in this case, of the city councils (and municipalities in general), towards the irrigation communities for the services they provide with their sustainable water management: (i) supply services: they provide water and food, mainly; (ii) regulation services: climate regulation, as they lower the temperature and increase the environmental humidity and regulation of hydrological cycles, they reduce soil erosion; (iii) support services, understood as an exceptional habitat for flora and fauna; (iv) cultural services, they are historical agrosystems, which can be offered as a sustainable tourism resource and learning for present solutions, coming from the past. In this sense, within the administrative agreements between the City Councils and the Irrigation Communities, one of the commitments on both sides has been the promotion of a network of cultural trails through the rural roads and the easement of the *acequias* managed by the Irrigation Community, combining their cultural and environmental function with the social and promoting their contribution to local economies.

The tool of payment for ecosystem services is undoubtedly innovative. It is true that there is already scientific literature on the subject, almost always from a theoretical perspective on the number and nature of these services or how to calculate their importance or value, or how to establish compensation or payment mechanisms. However, there are very few experiences that can serve as examples of good practice or even as learning from failed processes. For this reason, we have been obliged to act in a particularly prudent manner, always encouraging reflection and debate between the parties, and not being ambitious in the type of agreement so as not to force and create opposing situations. Our role has been one of mediation and facilitation, but both the town councils and the Irrigation Communities have asked us, as a university, to also be guarantors of these agreements.

The models that can be counted on are very few, if any, always related to environmental services. However, from our point of view, caution must always be taken against the danger of commodification of natural resources. In our case, payment for services does not necessarily refer to an economic payment, but to an exchange of services and mutual support between the administrative institutions and the local irrigation communities that implies a recognition by the former of the value and importance of the latter. We consider it essential not to lose sight of the local aspect and the contact and protection of the traditional knowledge that these communities have in terms of the historical management of natural resources, from which we must all learn.

## 10.5    The pilot case: cultural routes through historic irrigation systems

In the development of the routes, it has been essential for them to be approved by the Andalusian Mountaineering Federation as a form of protection, officialisation and dissemination. The approval process involves a series of steps in which different actors are involved (University, City Council and local communities), which makes it a slow process at times, but which brings a series of benefits: firstly, that the trails are official and that they have approved signage and signposting, which the whole population has assimilated and understands. In addition, all approved trails are included in official guides, which make them more widely known. And, of course, it is also compulsory for the promoting entity, in this case, the town councils, to maintain the path correctly.

The cultural routes along the *acequias* affect limited areas of intervention, with similar geographical characteristics and often limited water resources. However, each municipality offers very different elements and very specific casuistry. For the correct preparation of the itineraries and knowledge of the irrigation systems, as well as the subsequent publication of interpretative guides for the routes, a large amount of cartographic information has been compiled on all the *acequias* and the infrastructures related to them in each of the municipalities: main *acequias*, catchment areas, branches, mills, water mines, irrigation areas. All this is in order to generate a database of the hydraulic heritage of the area with which to obtain information on the

*Figure 10.3* Example of official signposting (official trail number: SL A 389) on the Acequia de Alcázar (Jérez del Marquesado. Granada)

*Figure 10.4* Mapping of *acequias* in Benamaurel (Granada)

evolution of the landscape, produce specific cartography for dissemination and use it as a tool for conservation and knowledge. A search for historical cartography and the collection of oral sources through interviews with members of the irrigation communities has also been carried out.

The town of Castril (see map) has been one of the pioneers in the proposal. It is a municipality where there are various water-related heritage sites used as tourist resources (Cerrada del río Castril trail), as well as a long history of fighting for the defence and protection of water resources, in this case, the river that bears the name of the town. The good relationship between the Irrigation Community and the Town Council has led to the signing of an administrative agreement between the two entities, not only for the execution and maintenance of the trail, which has been approved by the Andalusian Mountain Federation, but also to establish a more institutional link between the two entities, which will facilitate the resolution of conflicts and the provision of environmental services in the immediate future. Prior to the design of the trail, a historical-archaeological report was carried out on the irrigation community, compiling written testimonies about its historical existence, as well as the mapping of the entire network of *acequias* in the municipality. This was the precedent for the creation of an official, approved trail that would run through one of the most important areas of Castril, generated from these irrigation systems: the Vega de Tubos, which owes its name to the main spring, through which the route also runs.

Another example of good practice, not only in the creation of a path along the *acequias*, but also in the recovery of heritage and natural spaces has been the municipality of Jérez del Marquesado (see map). This town has a privileged location between the fertile plain and the northern foothills of the Sierra Nevada, which means that it has a large amount of water for a large part of the year. Three rivers run through the municipality: the Alhorí, the Bernal and the Alcázar. It is an ideal place in which five irrigation systems survive, which are supplied by a dense network of *acequias*. The origin of these systems is attributed to the medieval Islamic period, between the 8th and 10th centuries, but from the 12th century onwards, the five original villages were concentrated into a single one that makes up the current municipal territory.

Jérez del Marquesado is a municipality with which we have been collaborating for years. It is a territory subject to strong pressures for the technification and intensification of irrigation. As in other localities, there is a close collaboration between the Irrigation Community and the town council, although it is not without its problems. However, this relationship is by no means institutional, but rather governed by custom. In the case of small villages, most of the population owns land with water rights and is member of the Irrigation Community, including the political leaders of the local administration. The proposal of the agreements also serves to formalise and institutionalise these relations, giving them a legal framework.

Several heritage initiatives (Martos Rosillo et al., 2020) have formed part of the "Acequia de Alcázar" trail in the town of Jérez del Marquesado. It is a municipality with numerous heritage resources agricultural tradition that has shaped the landscape. An example of this is the area known as Alcázar, an area of great historical interest, which, in Islamic times, was one of the centres of population. Today, several remains of the settlement are still recognisable: the tower of Alcázar (14th century), the tower of Jérez and several remains of the old village, as well as the cultivation terraces and, of course, the *acequias*. The area around the Alcázar tower had been abandoned for about 50 years. In order to integrate this interesting place into the cultural route, its recovery was promoted, in collaboration with the Town Council, the Irrigation Community and the local Rambling Association: the excavation and restoration of the tower was carried out by the so-called "Escuela de Balates", which is the name given to the dry stone walls that support the cultivation terraces. The dry stone, declared Intangible Heritage of Humanity by UNESCO in 2018, was used to rebuild several of the terraces taught by several of the older residents of Jérez del Marquesado. The branch that irrigated these terraces, belonging to the Alcázar *acequias*, was also recovered and local fruit trees such as chestnut and rowan trees were planted. This initiative has not only achieved the recovery and investigation of the Alcázar neighbourhood, but also the participation and social involvement and the incorporation of all this space in the path along the Alcázar *acequias*.

Social involvement has also had two fundamental aspects: the participation of the population in the initiatives for the recovery of spaces and in the design of the routes, and the participation in the knowledge and dissemination of the routes. For this reason, a plan for the dissemination of the routes and their contents has been fundamental. Of course, all the tracks and descriptions of the trails have been stored on the Wikiloc platform, which has 12 million users and is translated into more than 25 languages. In order to diversify the spaces for dissemination and to link other research and proposals related to the subject, a specific section has been created for the routes along historic *acequias* on the website Regadío histórico (https://regadiohistorico.es/), a web infrastructure previously created to bring together collaborative information on irrigation systems in Granada and Almeria (Martín et al., 2022). The main objective of this section of the website is to host the current itineraries, as well as to guide for hikers. This complements the information, condensing in a didactic way all the research carried out on the irrigation systems of each municipality. The proposal for the future is to house more routes of this type, thus creating this concept of hiking routes along the path of the *acequias*.

Another way of dissemination has been through the Granada Tourist Board, a partner of the INCULTUM project and an organisation created for the promotion of tourism in Granada since 1982. Its website offers a wide variety of cultural offerings, among which are the Excursions and hiking in Granada|Patronato de Turismo.


## 10.6   Conclusions

The processes of design, creation, approval and dissemination of itineraries along historic *acequias* have been a tool for intervention in the territory, an alternative to mass tourism and the overexploitation of cultural and natural spaces. It has allowed us to learn more about the problems surrounding water, agriculture and heritage, which are so present today, as well as to propose solutions and alternatives through the mechanisms of payment for services.

Tourism, an action that is so present today, should be an intervention tool for the improvement of the territories and the people who live there. Our proposal integrates the basic pillars of a sustainable and regenerative tourism that ensures positive changes in the territory and the societies where it is developed. The aim of these itineraries is to promote sustainable economic development through the diversification of the cultural offering of the municipalities, boosting local consumption and the circular economy. It has promoted the social inclusion of marginalised and undervalued sectors of the population, such as Irrigation Communities. Under the trail proposal, the aim has been to conserve these resources, which form an essential part of the landscapes, as well as to preserve cultural values, diversity and heritage. The proposal is based on the multifunctionality of the agricultural areas, their cultural, environmental and agronomic values and the ecosystem services they provide.

From there, we have always tried to ensure that the process was participatory, but above all that it had a direct return for the Irrigation Communities themselves as a way not only of diversifying the economy and agricultural income, but also of distributing the benefits of tourism beyond the specialised economic sectors. In this case, the Irrigation Communities, the farmers, provide the whole municipality with a cultural service, for which they are paid through an administrative agreement with the town council. It is an apparently simple tool, but one that requires a great deal of mediation and facilitation work in order to reach satisfactory agreements for the parties. Above all, it is a tool that opens up a very interesting path for the communities and farmers themselves, not only as producers of food and raw materials, but also as service providers and, of course, custodians/caretakers of the territory.

Throughout the INCULTUM project, the interest that this type of resource, of paths along the *acequias*, could have among the population has been taken into account. A first initial survey showed that 98% of the people questioned were interested in this type of route. Also, a high percentage (93%) found it very interesting that the routes were guided by members of the Irrigation Communities and were willing to pay a small amount of money for it, thus contributing to the improvement of the economy of the communities. With this, we wanted to investigate the acceptance of this type of resource as another element that the Irrigation Communities could integrate into their organisational structure and that could be taken into account as a possible tourist resource, generating possible jobs.

So far, four itineraries along historic *acequias* have been carried out in the Altiplano region, with a total of 31 km of trails. This idea has aroused interest in other localities in the province of Granada, since, as we have been saying throughout this text, historical irrigation systems are very present in the geography of the southeast of the Iberian Peninsula and a fundamental part of rural society, but also of urban society. Two itineraries have been proposed in the region of La Alpujarra, which stands out for the particular water management systems known as "careos". The pilot municipalities have been Cáñar and Pórtugos, where the agreements for payment for services have been carried out for the elaboration of an official trail. A total of 17 km of trails have been signposted. Another initiative in the province, in this case in the capital of Granada, has been the recovery of one of the main *acequias* that supplied water to the capital in medieval times, the Aynadamar[3] *acequias*. Noy only has a section (3 km) of this canal, which had been abandoned for 50 years, has been recovered, but also the path next to it, to make it possible for the people of Granada to walk from the city along this now millenary system.

These trails along historic *acequias* should be understood as a tool for social intervention. A way of dynamising and diversifying the services provided by the local communities of these rural and, moreover, marginalised municipalities. Through an instrument as powerful as learning and education, channelled through sustainable tourism, alternatives can be proposed that benefit the rural population, that protect these socio-ecosystems and, in addition, generate social and environmental awareness. The objective is to generate a cultural service linked to the cultural landscape and agricultural heritage, built and managed by the irrigation communities. An infrastructure at the service of the local community and the municipality, not only for the enjoyment of the neighbours, but also as a tourist resource that allows diversifying the economy and complementing it and making the irrigation communities visible by informing the visiting public about LEK, practices and values.

## Notes

1 "Acequiero" is a person nominated by the irrigation community who is in charge of the distribution of water and also carries out social monitoring to avoid possible conflicts.
2 Documentary on the recovery of the Barjas irrigation channel in Cáñar (Granada) https://www.youtube.com/watch?v=TeQHef4NmI0.
3 Acequia de Aynadamar: the return of the fountain's tears https://www.youtube.com/watch?v=eg3WPPV45jo&t=66s.

## Bibliography

Alfaro Baena, Concepción. (1996–1997). "Castril: de Hisn frontero a señorío bajomedieval", *Anales de la Universidad de Alicante: Historia medieval* (Ejemplar dedicado a: Actas del Congreso Internacional "Jaime II, 700 años después"/coord. por Juan Antonio Barrio Barrio, José Vicente Cabezuelo Pliego, Juan Francisco Jiménez Alcázar), pp. 553–564.

Alfaro Baena, Concepción. (1998). *El Repartimiento de Castril. La formación de un señorío en el Reino de Granada*. Granada: Grupo de Investigación "Toponimia, Historia y Arqueología del Reino de Granada".

Aliod Sebastián, Ricarco, Corominas Masip, Joan, & Martínez Fernández, José. (2020). "Medidas para la transición a un regadío sostenible en el contexto del cambio climático y la DMA". En *XIX Congreso Ibérico de Gestión y Planificación del Agua. Transición hídrica y cambio global: del diagnóstico a la acción. 3–9 septiembre de 2020. Congreso Virtual.*

Brink, P. ten. (ed.) (2011). *TEEB – The Economics of Ecosystems and Biodiversity in National and International Policy Making.* London and Washington, DC: Earthscan.

Correa Jiménez, E., Abellán Santisteban, J., Martín Civantos, J. M., Román Punzón, J. M., & Bonet García, M. T. (on press). *Senderos culturales a través de acequias históricas en el altiplano granadino: el proyecto INCULTUM* en Sorroche Cuerva y Ruiz Álvarez. "ARQUITECTURA EXCAVADA Y PAISAJE CULTURAL" Editorial DYKINSON ISBN: 978-84-1122-901-2.

Delgado Anés, L. (2017). *Gestión, comunicación y participación social en los paisajes culturales de Andalucía: El caso del Proyecto MEMOLA.* Granada: Universidad de Granada.

Martín Civantos, José María. (2011). *The Archaeology of Irrigated Spaces in Southeast Spain During the Medieval Period.* Doi: 10.1484/M.RURALIA-EB.1. 100153.

Martín Civantos, J. M. (2014). "Mountainous landscape domestication. Management of non-cultivated productive areas in Sierra Nevada (Granada, Almería, Spain)". *Postclassical Archaeologies,* 4, pp. 99–130.

Martín Civantos, J. M., Toscano, M., Bonet García, M. T. y., & Correa Jiménez, E. (2022). "Un mapa colaborativo para documentar y difundir los sistemas de regadíos históricos de Granada y Almería". *Revista PH,* 105, pp. 12–14.

Martos Rosillo, S., Martín Civantos, J. M., Ramos Rodríguez, B., Abellán Santisteban, J., González Ramón, A., Jódar, J., Peinado Parra, T., Cifuentes, V. J., García Martínez, F. J., & Durán, J. J. (2020). "Recuperación de sistemas ancestrales de manejo del agua que utilizan soluciones basadas en la naturaleza. Las acequias de careo de Jérez del Marquesado". *XI Congreso Ibérico de Gestión y Planificación del Agua.* Septiembre 2020, 358–369.

ROMERO DÍAZ, M. (1989). Las Cuencas de los ríos Castril y Guardal (Cabecera del Guadalquivir). Estudio hidrogeomorfológico. Tesis Doctoral. Excmo Ayuntamiento de Huéscar (Granada) y Universidad de Murcia.

**Online references** (last connexion November 2023)

Altiplano de Granada. Grupo de Desarrollo Rural. https://altiplanogranada.org/

Digital meets Culture (9 February 2022). *Recovering abandoned spaces at the Dry Stone Balates School, Jérez del Marquesado (Granada).* https://www.digital-meetsculture.net/article/recovering-abandoned-spaces-at-the-dry-stone-balates-school-jerez-del-marquesado-granada/

Federación Andaluza de Montañismo. https://fedamon.es/

Geoparque de Granada. https://www.geoparquedegranada.com/

Laboratorio de Arqueología Biocultural (MEMOLab UGR). https://blogs.ugr.es/memolab/

Mapa colaborativo de regadíos históricos de Granada y Almería. https://regadiohistorico.es/

*Mediterranean Mountain Landscape.* https://memolaproject.eu/es

MEMOLAproject. (14 October 2014). *Documental sobre la recuperación de la acequia de Barjas en Cáñar (Granada)* [Video archive]. https://www.youtube.com/watch?v=TeQHef4NmI0

Patronato Provincial de Turismo de Granada. https://www.turgranada.es/patronato-provincial-de-turismo/

Secretaría General Universidad de Granada. *Presentación de la primera Asociación de Comunidades de Regantes Históricas y Tradicionales de Andalucía ACEQUIAS HISTÓRICAS* (19 April 2016). *https://secretariageneral.ugr.es/informacion/noticias/presentacion-de-la-primera-asociacion-de-comunidades-de-regantes-historicas-y-tradicionales-de-andalucia-acequias-historicas*

UGRmedia. (1 June 2023). *Acequia de Aynadamar: el regreso de las lágrimas de la fuente* [Video archive]. https://www.youtube.com/watch?v=eg3WPPV45jo

UNESCO. (2018). *Conocimientos y técnicas del arte de construir muros en piedra seca*. https://ich.unesco.org/es/RL/conocimientos-y-tecnicas-del-arte-de-construir-muros-en-piedra-seca-01393

# 11 Ancient paths to the future

## An integrated approach to territorial communities and entrepreneurship

*Flore Coppin and Vincent Guichard*

## 11.1 Introduction

The pilot project *Ancient Paths to the Future* (APF) is being tested in a small part of the Morvan Regional Natural Park (420 km², 3,800 inhabitants in 12 village communities), which covers a rural mountain region in Central France (3,000 km², 52,000 inhabitants) affected by a sharp fall in population (–80%) throughout the 20th century. Today, the area's economy is based on three pillars: cattle rearing, forestry and, increasingly over the last few decades, the residential economy (tourism and the arrival of new residents).

The area's tourist and residential appeal, which has increased since the COVID 19 crisis, is based on the recognised quality of its rural landscape and its relative proximity to major population centres, starting with the Paris Metropolitan area. However, the landscape is changing rapidly under the triple impact of the continuing decline in agricultural activity, brutal forestry practices and the already very visible impact of climate change on the forests.

Bibracte – Mont-Beuvray is the site of an ephemeral town dating from the 1st century BC and the local main attraction for tourists. It is protected as both a historical monument and a landscape heritage site. Since the 1990s it has been home to a major scientific and cultural facility managed by a dedicated public body. The Bibracte museum attracts almost 50,000 visitors a year, while the archaeological site attracts twice that number (Figure 11.2).

Based on the well-established visitor numbers to the Bibracte site, Ancient paths to the Future's (APF) ambition is to irrigate the surrounding area by developing a wide-ranging offer and improving the quality of services (accommodation, catering, mobility, etc.) by setting up a regional entrepreneurship scheme, encompassing the various sectors of activity that shape the landscape and the economy.

In addition, APF is concerned to develop a well-managed and community-based tourism offer that mobilises all the local players, including the professions that "make" the landscape. In other words, we believe that social empowerment and self-organisation are the only way to achieve a genuine

DOI: 10.4324/9781003422952-11

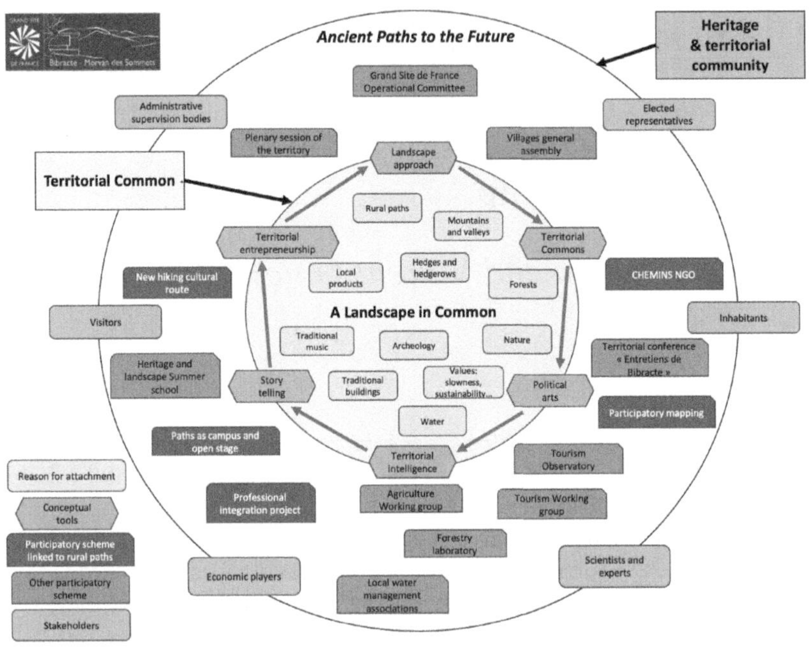



*Figure 11.1* Conceptual diagram of the territorial experimentation project into which the Ancient Paths to the Future pilot project fits. Copyright: Bibracte (Flore Coppin, Vincent Guichard).

enhancement of the area's heritage, that is not simply a "make-up for the tourist offer or the creation of a beautiful scenography for projects using the landscape as a showroom" (Poli, 2018, p. 110), and to ensure its economic development.

Figure 11.1 details the components of the project:

- A territory considered as a Common that is generating different reasons for attachment;
- Conceptual tools that are used in a logical order and in an iterative way, with each cycle helping to strengthen the approach a little more;
- Stakeholders who need to be involved;
- Participatory mechanisms developed to build an active heritage and territorial community of stakeholders.

The following paragraphs briefly describe the conceptual tools in the order in which they are used.

The regional project is based on a common territorial element, the landscape, which is expressed in a number of ways, such as the paths, the forest, the water and the material heritage bequeathed by the agro-pastoral society of past centuries.

The landscape approach is the driving force behind the community-building method, because of its ability to mobilise people and the holistic approach to the area that it enables.

The stakeholders involved in the project are organised into thematic consultation and project groups.

## 11.2   The landscape approach: cultivating an attachment to the landscape

### 11.2.1   *Expressing attachment to the landscape*

*Ancient Paths to the Future* assumes that an area can be promoted all the more effectively if its inhabitants are aware of its uniqueness and richness. This awareness is a factor in commitment and motivation of the community to maintain the area's heritage features. It fosters a sense of place attachment, collective spatial identity and pride.

Sharing the experience of attachment with others enables personal attachment to be extended to the dimensions of a community. The "symbolic" or "identity" appropriation of place is therefore associated with a social group to the point of becoming one of its attributes and to contribute to defining its social identity.

What's more, many local surveys show that place attachment is the most frequently cited source of well-being, and that this attachment is all the stronger if the area is unique in terms of its landscape, in other words, if it has escaped the trivialisation of space that affects the inhabitants of suburban and "urban" areas lacking in quality (Dissart and Seigneuret, 2020).

To achieve this, the project relies on the "landscape approach", understood as an

> Act of bringing order, coherence and a return to the common good in a fragmented territory, in a space that has hitherto been used by each discipline according to its own logic, for its own exclusive benefit and in deliberate ignorance of the specific character of the place. This unifying, all-encompassing approach obviously makes use of all the expert knowledge that deals with the many aspects of planning: from construction to agronomy, from ecological engineering to the art of building roads, from the know-how of botanists to the creativity of architects.
>
> (Thibault, 2022, p. 40)

*Ancient Paths to the Future* is part of the *Grand Site de France* approach, which has been implemented at Bibracte since the 2000s. This approach stems from the French government's concern, expressed in the late 1970s, to combat the harmful effects of overtourism on the country's most emblematic protected natural and heritage sites. *Grand Site de France* is a label awarded by the French ministry responsible for landscape policy, which is reviewed every six years. The label is designed to encourage local authorities to take responsibility for the sustainable management of their most emblematic sites, going well beyond tourism issues alone. One of the special features of this policy is that it is not prescriptive: applicants are free to define the geographical scope of the designated area, mobilise stakeholders and organise governance. In this sense, the *Grand Site* approach is a genuine laboratory for innovation in the management of rural areas. It assumes that the landscape approach is an effective and virtuous lever for territorial action, through its capacity to mobilise stakeholders and its ability to encourage a holistic and integrated approach to the territorial project. Some 50 local authorities are involved in the approach; they are federated within an independent organisation, the *Réseau des Grands Sites de France*, which over the years has become a think tank for public policies based on the landscape approach and which shares its experience within an international training centre (www.grandsitedefrance.com/en/) (Boisseaux et al., 2022).

The integrated territorial project developed around the Bibracte site took on a new dimension in 2021 when the *Grand Site de France* label was renewed for the second time. The public body managing the heritage site is the lead partner, with two other public partners' organisations, the Morvan Regional Nature Park and the Nièvre District. The project is resolutely experimental. Its means of action mainly involve support for research and innovation: at European level, the H2020/Horizon Europe programmes and the European Innovation Partnership for Agriculture, and at national level, calls for projects from the French National Research Agency.

A large multidisciplinary scientific community is being mobilised across the region. In a rural context, where the population is sociologically very fragmented, we are also banking on artistic mediation to establish dialogue, to renew the way in which local people look at their environment and to help them share in the "storytelling" of the regional project. Since 2021, a cultural coordinator and artists have been working on this project thanks to the European LEADER programme.

At the local level, the mayors of the 12 village communities involved have been given a special role, while thematic working groups have been set up to explore the various issues facing the area. A number of bodies with legal status have also been set up at the same time. An association provides support for the Agriculture working group, while another, CHEMINS, is a much more cross-disciplinary non-profit association working on one of the area's most important heritage themes.

The approach is run on a day-to-day basis by a small operational team made up of staff from Bibracte and its partners, some of whom are paid on a project basis. As part of this team, the INCULTUM project manager led the APF project.

## 11.3   Recognising the landscape as a common good

As defined by the European Landscape Convention, landscape is considered to be a part of a territory as perceived by the population, whose character is the result of the action of natural or human factors and their interrelationships. As a resource that contributes to the general interest, the management of which implies responsibilities for everyone, the landscape defined in this way can be considered as a common good in the sense of Elinor Ostrom (1990), as long as it is considered as a resource that gives rise to a community's attachment to its living environment, and as long as the community provides itself with the means to manage it using appropriate rules.

Considering a landscape as an ecosystem, an area inhabited by a wide range of socio-economic players and an object of shared attachment (Besse, 2018) also makes it possible to address all the diverse issues facing an area in an integrated way: agriculture and forestry, water and biodiversity, tourism and the residential economy and, of course, local governance. The success of the approach requires a collective awareness of the territory, understood as "the awareness acquired through a process of cultural transformation by the inhabitants, of the heritage value, of the territorial common goods, material and relational, as essential elements for the reproduction of individual and collective, biological and cultural life", to use the vocabulary of the work of the Italian territorialist school (Magnaghi, 2010). In the terminology of the *Grands Sites de France*, the term "*spirit of the place*", genius loci, is used to express this concern metaphorically. From an operational point of view, it is a question of identifying the constituent elements of the territory's experience that provoke the shared attachment of its inhabitants. In the case of Bibracte, the area's exceptionally dense network of paths proved to be an excellent indicator of this attachment.

## 11.4   Rural paths: the revelation of a common ground

For the territorialist school, "representing the different specialised knowledge of territorial heritage is the first step towards knowing, managing and socially reproducing territorial heritage", in order to "re-territorialise" policies that throughout the 20th century led to territories being reduced to spaces to be developed devoid of identity (Poli, 2018, p. 117).

With this in mind, the Grands Sites de France are developing participatory tools and methods that aim at giving everyone a voice in order to bring out the reasons for attachment shared by the inhabitants, which define the area's heritage value. This heritage value forms the basis of the regional project, which aims to place the development of this area within a long-term historical trajectory while respecting its heritage value.

In the case of Bibracte, preparations for the 2nd renewal of the *Grand Site de France* label, between 2017 and 2021, have made it possible to redefine

*Figure 11.2* Restoration work on an abandoned path on the Bibracte – Morvan des Sommets Grand Site de France. Copyright: Bibracte (Antoine Maillier).

the boundaries of this area and the elements that make up its heritage value. In practical terms, this survey took the form of Landscape and Heritage Days organised in each village community. On each occasion, local councillors and interested residents were invited to show us, along a walking route, the places they found most interesting in their environment. These days inevitably revealed a strong attachment to the paths.

In fact, the existence of a very dense network of footpaths is a unique feature of the Morvan. This network is the result of a combination of two factors: the wide dispersal of settlements since the Middle Ages and the lack of agricultural consolidation in the second half of the 20th century. With the population declining by 80% over the last century, many of these paths are now disused and buried under bushes and woodland, but almost all of them have been preserved. Even more interestingly, they are legally rural roads, owned by the village communities as a public authority.

This network was largely underestimated by the inhabitants and their elected representatives and its legal status was also poorly understood. It is the vestige of a genuine, centuries-old territorial common that has been neglected since the beginning of the rural exodus at the end of the 19th century. What's more, attachment to the paths is largely based on the quality of the landscapes they cross. Preserving the paths then naturally leads to concern for preserving the quality of the countryside as a whole. This attachment, which was not foreseen by the initiators of the approach, now serves as a powerful lever for building an active heritage community that includes the 12 village communities concerned.

The network of paths criss-crosses the landscape, providing a framework and hierarchy for the landscape features, showing how the village land is appropriated; it is a spatial and social revealer. The path is also characterised by an associated "small heritage": dry stone walls, stream crossings, fountains and calvaries; it bears witness to ancient agricultural know-how and the rational use of local resources: planted hedges, fruit trees, spring catchments, etc. The path also provides environmental services: its hedgerows protect against erosion, enrich the soil, improve water quality and moderate the climate; an essential link in the preservation of biodiversity, they offer refuge to many species and act as an ecological corridor. The path offers a range of amenities; it is a much sought-after and much-appreciated place for walking. Its maintenance is justified by different, sometimes conflicting, uses: access to dwellings or agricultural plots, logging, recreational or sporting use (hunting, hiking, etc.).

It is also worth noting that this new interest in footpaths in France is part of a broad national movement, reflected in numerous local initiatives.

## 11.5　The heritage inventory, a first step towards the "territorialisation" process

The internship of a geography and geomatics university student has made it possible to develop and validate a protocol for inventorying and classifying the paths, using a GIS embedded in a tablet (QFIELD) so that the data can be entered in the field: the actual route of the path and associated parameters (heritage value, uses, state of conservation, maintenance methods). This inventory is based on a network of referents identified in the town councils and brought together in a dedicated working group, with the *Grand Site de France* management team providing the method and technical advice.

The inventory resulted in an atlas of the region's rural footpaths, with a network of 1,100 km covering an area of 420 km². This heritage is all the more important given that, in the Morvan, village communities rarely own any other communal land.

Only a third of these paths are currently the subject of formal protection (included in departmental hiking policy) and management measures, with dedicated resources provided either by the village communities (*Communes*) or by groups of village communities (*Communautés de communes*). For the others, maintenance is left to the initiative of the users (inhabitants, farmers, foresters, hunters, sports organisations, etc.) according to their needs.

During the summer of 2022, a "General Assembly on Rural Paths" brought together most of the parties involved in their management – villages, groups of village communities, districts (*Départements*), tourist offices, the French hiking federation, etc. – and reiterated the particularly complex regulatory and administrative framework for their management.

Following consultation at the end of 2022, the village communities then embarked on a shared approach to path management. One of the aims of this approach is to produce a rural path management plan in the short term, which could take the form of a charter.

## 11.6   Levers for activating and stimulating the heritage community

### 11.6.1   *Supporting the emergence of the heritage community*

Historically, the collective management of assets of community interest such as paths has been slowed down, and often even interrupted, by several consecutive events. In our case, the first was the promulgation of the Civil Code following the French Revolution of 1789, which enshrined the right of ownership (private and public) and swept aside the notion of community resources. Then came the collapse of traditional peasant society throughout the 20th century. At the start of the 21st century, community management of resources in rural areas has become marginal – in the Morvan, it still exists only for some of the sources of drinking water – and we need to reinvent it in a socio-economic framework that is radically different from that of traditional peasant society.

Reactivating the common good involves raising collective awareness of the landscape resource – which has been achieved over the course of the INCULTUM project – and then establishing management rules and mobilising sustainable resources to ensure its preservation. The joint creation of a network of paths that criss-cross the landscape is thus a step towards the joint creation of the landscape as a whole.

Once the network of paths had been identified as a key factor in attachment to the area, the next step in activating this attachment was to explore the paths with the help of landscape illustrators working in the frame of artistic residencies. This group exploration provided an opportunity to gather everyone's point of view on the area's landscapes and to collect the stories associated with the area. The landscape artists capture the feelings of the participants and translate them into images and writing in the form of village survey notebooks, which evoke in a poetic and sensitive way the spaces and landscapes travelled through, experienced, loved, looked at and admired by the residents along the paths. Produced in small runs using a traditional silk-screen printing process, but also available for consultation on the *Grand Site de France* website, these notebooks make a major contribution to strengthening the attachment of local residents. They also help to produce original images that are a favourite material for the area's communication tools.

### 11.6.2   *Organising the heritage community*

As part of the *Grand Site de France* initiative, the activation of the landscape community is based on working groups formed around the different components and issues of the landscape.

These groups are led by members of the management team, supported as far as possible by scientists, experts and cultural players involved in the territorial project.

The Rural Paths group brings together around 30 people (elected representatives, volunteer residents and local professionals). Its main objectives are to:

- Draw up an inventory of the network of footpaths and their uses;
- Prepare a shared management plan based on priority uses;
- In the area of recreational and sporting uses, to identify routes, enhance them (by restoring them and marking them out), organise the services required by users (access, accommodation, catering) and promote them (publication of maps);
- Work in synergy with other working groups where necessary.

The Rural Paths group is led by the GSF's Heritage and Tourism Officer, who is also the APF project coordinator. Her role is to help set and achieve concrete short-term objectives, which is essential to ensure the long-term commitment of the group's members. This involves drawing up the inventory (atlas), putting in place tools and working methods (field survey, maps, etc.), organising field trips and participative workcamps and devising and implementing an enhancement strategy.

The assessment of the current situation and the preparation of the management plan will primarily involve elected representatives and members of the public who are members of the working group, while the strengthening of recreational and sporting activities will also involve tourism and service providers, in particular those who are members of the Tourism group.

### 11.6.3   *Enhancing the value of the path network for local residents and stakeholders*

Different approaches are being used to enhance the value of the network of paths and strengthen the attachment of local residents to it. We have already mentioned the artistic survey notebooks, which played a decisive role in the initial phase of the project. We're continuing here with two other actions: the organisation of enhancement projects and cultural walks along the paths.

In 2022 and 2023, volunteer summer heritage work camps were organised in partnership with the national NGO *Rempart* Federation, which brings together nearly 200 associations that organise heritage restoration work camps open to volunteers, and the regional association *Tremplin – Hommes et Patrimoine*, which uses heritage restoration as a means of socio-economic reintegration. The dozen or so volunteers were able to take part in the restoration of an abandoned path and its dry-stone boundary walls and to learn about the Grand Site de France approach and the regional project developed within this framework, thanks to contributions from experts (archaeologist, restorer, landscape architect, etc.). This original offer has attracted architecture, archaeology, landscape and art history students who are sensitive to environmental issues. As a result of the work carried out, a section of the path has been restored and is now part of a new discovery trail on the slopes of Mont Beuvray. At the end of this experiment, the aim is to make this training course a permanent feature in the form of a summer school, during which students will come and learn about the working method used in the *Bibracte – Morvan des Sommets* ecological transition laboratory, while also getting their "hands dirty".

Walking along the paths is a great way of discovering the richness of the area and, at the same time, creating links between those taking part in the excursions. For this reason, as part of the pilot project, cultural walks have been organised along the paths under the name of *Balades attentionnées* – the word *attention* being used in French in two complementary senses: the concentration required of participants in order to be attentive to their environment and the solicitude required to preserve that environment. In addition to providing an opportunity to discover the area along the way, these walks showcase the people involved in the area (a farmer who looks after his hedges in an exemplary way, the owner of a water mill who has restored it, etc.). They also involve "experts" (such as scientists invited to explain the area's geology, hydrology and biodiversity, environmental and forest management technicians, etc.) and artists who are invited to "shift the focus" of participants and facilitate dialogue between them. In 2022–2023, around 15 walks were organised in the 12 villages of the area, each time attracting several dozen participants, which is more than enough to manage the group and encourage discussion. It should also be noted that participants were attracted mainly by word-of-mouth, without any formal (and costly) means of promotion.

Through these actions, the paths are regaining their function as spaces for sociability and are tending to become third places, understood as

> spaces in which the desire of a community of citizens to move towards a better world is embodied, whose network redraws the territory in which they are anchored with common sense, cooperation and solidarity, and positions itself at the heart of exchanges between public players, private players and citizens.
>
> (Wikipedia France, adapted)

From another point of view, we could also talk about Political Arts practical workshops along the way. In the thinking of sociologist and philosopher Bruno Latour, Political Arts consists of the simultaneous use of scientific methods and artistic practices to analyse societal issues and enrich political decision-making processes (Latour, 2021).

### 11.6.4   *Taking a step back and learning from the experience of others*

Organised on an annual basis, the conference *Entretiens de Bibracte-Morvan* was initially intended to discuss the issues facing the region in the light of scientific opinion. Open to all, in recent years, they have explored the concepts of the Commons, climate change, solidarity between humans and non-humans and agro-ecology. The concepts were explored in practical terms through sessions organised in the field, in close contact with local players and, for the most recent editions, with the involvement of artists.

In 2023, the 17th *Entretiens*, organised as part of the INCULTUM pilot project, looked at the concept of heritage and the role of the arts and cultural action in fostering local people's attachment to their area. Over 150 participants took part in the series of lectures, walks and meetings, sharing experiences and engaging debates.

## 11.7   The tourism project as a lever for the regional project

### 11.7.1   *Encouraging and supporting the projects that the region needs*

Elinor Ostrom (1990) gave a new perspective to the management of common goods by giving them a value of utility and the creation of a civil and solidarity-based economy. The *Grands Sites de France* landscape approach, which calls on citizens to manage and, where necessary, transform a common asset to facilitate the ecological transition, is in line with this principle.

In this respect, the development of territorial entrepreneurship is an integral part of the territorial approach. In France, territorial entrepreneurship is defined by the *Banque des Territoires* (a large French State-owned bank), as "an entrepreneurial movement that reinvents new, more collective ways of doing business, with the aim of generating responses in favour of economic development that is more locally rooted, more sustainable and more inclusive". It can cover a wide range of issues: safeguarding traditional activities, maintaining the rural socio-economic fabric, adding value to local products, developing new activities, etc. It is an alternative between public action and private entrepreneurial projects. This type of entrepreneurship can take the form of a traditional business or new forms of cooperatives based on the social economy.

Responsibility for planning, financing and managing projects is left to local players and legitimate local authorities, with the *Grand Site de France* community offering socio-economic players a space conducive to dialogue and project construction, with the support of the *Grand Site de France* initiative management team, which is striving to integrate the project management dimension of territorial entrepreneurship, which consists of:

- Federating, decompartmentalising and guiding professionals in their desire to bring projects to life by promoting partnerships;
- Leading multi-sectoral working groups and interest groups in the development of the tourism offer;
- Helping to (re)discover local industries by promoting them;
- Supporting new entrepreneurs by encouraging multi-activity, particularly between the agricultural and service sectors, in order to strengthen the economic viability of projects;

- Developing skills;
- Acting as a mediator and ensuring that the project is fair, so that no local resident feels left out.

### 11.7.2  *A federative initiative: the* Tour du Morvan des Sommets

The *Tour du Morvan des Sommets* is a new cultural itinerary for discovering the region and the flagship project developed as part of INCULTUM. It responds to a desire shared by the elected representatives of the 12 village communities and its implementation makes a major contribution to the desire to work together on a regional scale.

Workshops held by the Rural Paths group to design the route led to the definition of an itinerary that has been improved with the support of local walking associations and technicians from the Morvan Regional Natural Park.

The *Tour* is a 140 km long hiking tour that links the 12 villages, making the best possible use of the network of paths. By linking the 12 village communities, the itinerary invites visitors to discover a rich and living heritage that expresses the ways in which people have lived here, past and present. It aims to become a tool for raising visitors' awareness of climate change and a showroom for traditional practices and know-how that can inspire solutions for sustainable land management.

The route was tested in 2022 by a tourism professional, who validated its interest and feasibility. It was put into service in summer of 2023, with the support of temporary promotional material pending its homologation as a national hiking route, which will enable it to appear on the maps published by the French National Institute of Geographic and Forest Information.

### 11.7.3  *Sectoral working groups on the common landscape*

This discovery route now serves as a catalyst for landscape initiatives and is a means of federating economic activities based on shared landscape resources.

In addition to the Rural Paths group mentioned above, the *Grand Site de France* initiative gathers four other thematic groups.

The Tourism group brings together around 20 local players (tourist offices and tourism service providers) who are working to build a "slow tourism" offer throughout the four seasons, combining leisure activities and outdoor sports, heritage and cultural discovery and encounters with local people, particularly along the *Tour du Morvan des Sommets*.

The Art & Territory group brings together around 20 local artists to develop collective projects.

The Agriculture group was set up to forge links between long-established farmers, all of whom raise cattle, and new farmers, who are often developing diversified projects geared towards short distribution channels. To this end, it has the support of the *Grand Site de France* project manager, who is seconded on a part-time basis by a local chamber of agriculture. The priorities are to facilitate the takeover of farms, create links with the local economy (farm visits, farm-to-fork offers) and strengthen solidarity within the farmers community. The group is now an association set up in 2022 with around ten members and it has been accredited as *a Groupement d'intérêt économique et Environnemental* (GIEE), a Ministry of Agriculture label designed to promote the local organisation of players in the agricultural sector around sustainable collective projects. The next step is to implement concrete projects, such as the creation of a reception area or the development of a catering offer available in the form of buffets.

The Forest group is a part of the forest experimentation laboratory set up in 2021. Led by Bibracte's Forest Project Manager, this group is being set up at the time of writing with the aim of establishing a dialogue on forest management, at a time when it is the subject of very lively debate in the Morvan, because industrial logging methods are damaging the landscape in a way that is less and less accepted by the local population.

Another example is the participatory approach launched in autumn 2023 on the subject of water, designed both to better characterise this resource, which is subject to shortages in the summer months, and to make it a component of the local community by relying on an original system of shared management of springs, the *Associations syndicales libres*, which is widely developed locally. Here again, the creation of a dedicated working group is a means of action that we aim to deploy.

Finally, the theme of professional integration has been developed over the last two decades to help enhance the Bibracte – Mont-Beuvray heritage site (restoration of archaeological remains, forestry maintenance). It is now being mobilised in support of the *Grand Site de France* initiative, for example to maintain the network of paths.


### 11.7.4   *The region's tourism observatory*

The development of the tourist economy is a real lever for the sustainable development of the region. However, this development potential comes up against widespread mistrust among the local population, who fear that their living environment is being "touristified" in an uncontrolled way, as other areas are suffering.

In response to this mistrust, a system designed to create a form of territorial tourism intelligence has been in place at Bibracte since 2019. Based on the EVALTO method (Fabry et al., 2012), the aim is to overcome preconceived ideas and objectify knowledge of tourism activity by means of in-depth surveys produced and analysed with local residents and stakeholders. In fact, the first edition of the survey showed that both local decision-makers and local residents had a very distorted perception of tourism, particularly as regards holiday tourism, which is much more developed than they think. A new edition of the survey will include hiking and green tourism around Bibracte, in order to provide a more complete picture of the profile and practices of visitors to the area.

## 11.8  Conclusion

By using sustainable cultural tourism as a means of recognising and diversifying economic activities based on the resources of a shared landscape, *Ancient paths to the future* is a local initiative that strengthens territorial synergies and enables tourism policy to be seen not just as a policy of economic and residential attractiveness, but as one facet of an integrated territorial project. This territorial experiment is an attempt to put into practice the recommendations of Bruno Latour, host of the *Entretiens de Bibracte-Morvan* in 2019, for whom it is vital and urgent to "land", i.e. to invent on a local scale, in each place, the arrangements that will enable the community of the living to preserve the habitability of the Planet at a time when it is seriously endangered by the entry into the Anthropocene (Latour, 2018).

The pilot project has shown that a local community can develop as a local "collective enterprise" in which all the players have a role to play, provided that a shared vision of the territorial project has been built collectively through the activation of a territorial common. This common ground is made up of the various elements of the area that are a source of attachment for its residents and visitors and which, taken as a whole, make up its landscape.

The three years of the *Ancient Paths to the Future* project, which were also the time when the Grand Site de France Bibracte – Morvan des Sommets project was gaining momentum, will obviously not have been enough to put in place a sustainable entrepreneurship system capable of ensuring the economic viability of the local project, but this prospect seems to us to be much more attainable at the end of the project, now that the mobilisation of local players has been achieved. Its success will clearly depend on the ability of Bibracte and its partners to strengthen the synergies between the various categories of players in the interests of a shared project, which will undoubtedly involve, as a priority, decompartmentalising the area's three main sectors of economic activity (agriculture, forestry and services) and creating multi-skilled professional profiles straddling the three sectors. This will bring us closer to the characteristics of the rural economy that have been erased by the specialisation that has been promoted since the post-war decades in order to increase the productivity of rural areas. The renaissance of multi-activity, backed up by a concern to preserve the territorial resources on which people rely, is undoubtedly a factor of resilience that will enable us to better face the many challenges posed such as climate change or depopulation.

### Acknowledgements

This chapter is the result of a collective process, and it would take too long to list all the players involved: elected representatives, residents, professionals, scientists, experts, etc. We would like to thank them all for their involvement and apologise for not mentioning them by name.

## Bibliography

websites: www.bibracte.fr/en, www.grandsitedefrance.com/en/, https://grandsite-bibracte-morvan.fr/

Besse, J.-M. (2018). *La Nécessité du paysage*. Paris: Parenthèses.

Boisseaux, T., Baccaïni, B., & Cabrit, J.-L. (2022). *Les Grands Sites de France, la force fédératrice du paysage au cœur de l'action des territoires*. Paris: Inspection générale de l'Environnement et du Développement durable (Rapport n° 014170-01). https://www.igedd.developpement-durable.gouv.fr/IMG/pdf/014170-01_rapport_publie_cle5b9942.pdf

Dissart, J.-C., & Seigneuret, N. (2020). Introduction. In: Dissart, J.-Chr., & Seigneuret, N. (Eds.), *Local sources, territorial development and well-being* (pp. 1–13). Cheltenham: Edward Elgar. https://doi.org/10.4337/9781789908619

Fabry, N., Zeghni, S., & Martinetti, J.-P. (2012). L'innovation soutenable dans le tourisme : le cas de la Cité Européenne de la Culture et du Tourisme Durable (CECTD). *Management & Avenir*, 56, 100–113. https://doi.org/10.3917/mav.056.0100

Latour, B. (2018). *Down to Earth: Politics in the new climatic regime?* Cambridge: Polity. (translation of: Atterrir. Paris: La Découverte, 2017).

Latour, B. (2021). Comment les arts peuvent-ils nous aider à réagir à la crise politique et climatique? *L'Observatoire*, 57, 23–26. https://doi.org/10.3917/lobs.057.0023

Magnaghi, A. (2010). *Il progetto locale. Verso la coscienza di luogo* (2nd edition). Torino: Bollati Boringhieri.

Ostrom, E. (1990). *Governing the commons. The evolution of institutions for collective action*. Cambridge: Cambridge University Press. https://doi.org/10.1017/CBO9780511807763

Poli, D. (2018). *Formes et figures du projet local: la patrimonialisation contemporaine du territoire*. Paris: Eteropia.

Thibault, J.-P. (2022). *Aménager les territoires du bien-être*. Paris: Le Moniteur.

# 12 Measuring rural tourist behaviour and engagement

## Inside the mind of the visitor

*Sabine Gebert Persson, John Östh,
Mikael Gidhagen, Marina Toger and
Anna-Carin Nordvall*

### 12.1 Introduction

Rural tourism has been emphasized as a sector highly prioritized by a number of countries, as it offers a means to stimulate socio-economic development by generating new job opportunities and providing alternative income streams for rural communities (Quaranta et al., 2016; UNWTO, 2023). Within hospitality research, customer engagement is identified as an important factor that can "enhance the overall visitor experience as well as the value proposition of the destination" (Bergel & Brock, 2019, p. 576). A challenge though is the double-edged sword of attracting tourists while preserving cultural heritage and minimizing the negative effects associated with tourism (Cöster et al., 2023), thus ensuring sustainable development of destinations. To achieve a balance, it is important to consider the specific context of the rural destination in relation to what attracts and engages visitors. When an individual is engaged in an object – a destination – this demonstrates their inclination to dedicate resources towards learning more about the place, interacting and discussing their experiences with others, or expressing their opinions (so-called word of mouth, WOM) (Harrigan et al., 2017). Previous research has demonstrated the positive effects of visitor engagement on revisits, commitment, loyalty, spending, and on relationships between the visitor and different stakeholders at a destination (cf. Rather et al., 2023). Hence, understanding the underlying processes and mechanisms driving visitor engagement, including the triggers of engagement, is central for attracting visitors to a destination, especially in rural areas. Given that the experience of a visit unfolds within a spatio-temporal context, real-time measurement of visitors' behaviour is necessary for understanding the factors engaging tourists (De Cantis et al., 2016). While visitor engagement is evolving through on-site experiences such as activities, encounters with different services, and interactions with the local communities and the landscape, previous studies on tourist engagement rely on post hoc survey-based data (So et al., 2014). Although survey data offer insights into preferences and motives, the results capture the cognitive experience perceptions *after* the visit, i.e. post hoc,

DOI: 10.4324/9781003422952-12

rather than the mid-visit experiences. Therefore, while survey data can help researchers and practitioners in understanding engagement *post*-visit, it fails to provide insights into the engagement process occurring during spatio-temporal mid-visit engagement. One way of tracking real-time movement has emerged with new technology, such as GPS loggers (De Cantis et al., 2016; Ferrante et al., 2018). However, while capturing movements, such technology has as of yet not been utilized for measuring tourist engagement. This chapter contributes to research on visitor engagement with a method combining GPS trajectories, OpenStreetMap, and surveys to capture and measure visitor engagement and experienced value. By illustrating how the method has been implemented in rural destinations, the research contributes to the understanding of the nature of spatio-temporal engagement in a destination and provides insights into how engagement affects individuals' value co-creation processes. In doing so, it contributes to research on engagement (Bergel & Brock, 2019; Brodie et al., 2019; Dessart et al., 2016), while offering practical insights for destination developers in rural areas on how engagement levels are formed and what development efforts should focus on for further growth or in mitigating overcrowding risks. Here, "visitors" refers to consumers engaging in tourism activities (Smith, 2000) and travelling to places outside their usual environment for no more than a year (cf. Frechtling, 2006).

## 12.2   Engagement and experienced value

Given our interest in understanding the mechanisms underlying engagement, it is essential to start with defining the concept per se. When an individual is engaged in an object or, as in this case, a destination, it reflects that individual's disposition to invest resources into learning more about the destination, expressing emotions, and sharing experiences (Harrigan et al., 2017). Previous research shows that engagement has a positive effect on loyalty, intention to return, and brand evaluations (cf. Harrigan et al., 2017; Kumar et al., 2010; Kumar & Pansari, 2016). Understanding how visitors become engaged in a rural destination, and in preserving the cultural heritage, requires measurement of the process of engagement. This helps to identify what triggers engagement and to track its evolution over time, i.e. both prior to, during (also called mid-visit), and after the visit. Despite the acknowledged importance of engagement in marketing research, there is a lack of consensus regarding its conceptualization (cf. Rather et al., 2023). Within engagement research, there are two dominant streams, which view it either as behavioural (cf. Harmeling et al., 2017; Kumar et al., 2010), or as a psychological state – sometimes also referred to as the disposition to engage (cf. Brodie et al., 2011; Rather et al., 2023; Storbacka, 2019). Within the behavioural approach, customer engagement is defined by behavioural manifestations, such as customers posting reviews, generating content concerning a product, making suggestions for improvements, or through spreading information by

WOM. Common to all these forms is that they to different extents add value to the firm/object, as the customer's activities indirectly or directly contribute to strengthening the brand (cf. van Doorn et al., 2010). The customer in this view becomes a marketer of the firm, brand, or object. All of these activities are voluntary from the customer side and are based on the willingness of the customer to promote a specific brand or firm. Defining engagement by behavioural manifestations suggests that it is a unidimensional concept, focusing on behaviour that goes beyond a core economic transaction (Harmeling et al., 2017).

While the unidimensional perspective of engagement contributes with insights on how engagement behaviour can be captured, it has been criticized for disregarding the cognitive and affective/emotional processes, i.e. the psychological state, which plays a part in forming and influencing customer engagement behaviour (cf. Brodie et al., 2011). By including the cognitive and affective/emotional processes, not only the manifestations but also the dispositions to engage (Storbacka, 2019) are involved. Combining the psychological state with the behavioural view offers possibilities for capturing the whole process of engagement, from the disposition to engage to the manifestations expressing the engagement. Considering that tourism is an experiential activity in which hedonic consumption is an essential part (experiencing feelings such as fun, excitement, or curiosity), both the behavioural aspects and the psychological state are important for understanding visitor engagement (Zhang et al., 2018). This multidimensional perspective has been defined by Brodie et al. (2011, p. 260) as:

> a psychological state that occurs by virtue of interactive, cocreative customer experiences with a focal agent/object (e.g., a brand) in focal service relationships. It exists as a dynamic, iterative process within service relationships that cocreate value. It is a multidimensional concept subject to a context- and/or stakeholder-specific expression of relevant cognitive, emotional and/or behavioral dimensions.

This quote highlights important aspects of the multidimensionality of engagement; its dynamic nature, and how it is dependent on the context in which the interactions take place. This dynamic nature furthermore implies that the level of engagement changes and depends on an individual's perceptions and reactions to an experience. The cognitive aspect of engagement denotes the mental processes in relation to an object and refers to how an individual activates their cognitive processes, such as seeking information. It also encompasses the willingness to exert that effort (Brodie et al., 2011; Hollebeek et al., 2019). The cognitive aspects thus include the mental processes and investments in an object, such as a destination. Whereas the cognitive aspect involves a process that requires the mental effort of an actor, the affective aspect is an emotional reaction to something, or a feeling (Dessart et al.,

2016; Hollebeek et al., 2019). This feeling is also something that destinations appeal to when promoting love, passion, or other emotional expressions. To summarize, the cognitive and affective/emotional aspects are an individual's tendency, here called disposition, to engage in an object (Fehrer et al., 2018; Storbacka et al., 2016). It is the continuous process of adjustments of the disposition that has effects on and is revealed through engagement behaviour. In this chapter, we apply a multidimensional approach to engagement, as it allows the researcher to capture both the enduring psychological connection to a brand, object, or destination in conjunction with behavioural manifestations (So et al., 2014). While studying activities may provide an indication of the individual's behaviours, the psychological state provides an understanding of the engagement valence, where positive engagement is an actor's positively associated feelings or evaluations towards an object, brand, or destination. This positive engagement is important for the "enhanced value co-creation experience" and is expressed in the form of behaviours such as repurchasing, positive WOM, and/or collaborations (Li et al., 2018, p. 492).

### 12.2.1   *Spatio-temporal context of actor engagement*

While a visitor's experiences during a visit are crucial for shaping their disposition and willingness to become engaged in the destination, the experiences and psychological expressions are embedded in and affected by the time and place in which they occur. The experiences are influenced by the context through a variation of touchpoints or interactions with other actors or the place, as "each and every actor experience occurs in a specific time and place, the connections surrounding the experience contribute to the framing of a psychological state or disposition" (Chandler & Lusch, 2015, p. 9). Thus, these connections, in a spatio-temporal context, in turn influence and are influenced by the cognitive and affective processes as well as the valence and strength of the visitor's engagement. Through a number of touchpoints or interactions, taking place before, during, or after the visit, the experiences and valuations of the destination and the visit evolve. Interactions are here understood as interactions with the place and also with "social others" (i.e., travelling companions), individually or in a group, which are essential in forming and influencing an individual's visit (Hamilton et al., 2021) during the entire process. Prior to the visit, the individual may have different reasons to go to a particular destination and a varying level of prior knowledge. This can in turn result in a search for information online, or by asking friends, family, or other individuals. The degree of attention, vigour, and enthusiasm towards learning more about the place represents the individual's affective and cognitive dimensions of engagement pre-visit. Engagement valence can be positively or negatively influenced or can remain unchanged. As the individual, alone or with others, visits and interacts with the destination (and its historical artefacts, traditions, languages, architecture, artistic expressions, inhabitants, and other visitors), the affective and cognitive aspects are evaluated and interpreted directly or indirectly. Through the interactive processes, the experiences are in turn formed, reformed, and reinterpreted mid-visit. As the visit ends, the value co-created through impressions and experiences have effects on the disposition to engage. Recognizing its multidimensional nature, as highlighted by for example Brodie et al. (2011), that evolves over time and in a context, implies that measuring engagement post hoc is insufficient for capturing the intricacies of both the engagement process and its multidimensionality. Following this line of reasoning, we argue for the necessity of measuring pertinent indicators of engagement throughout the whole process.

### 12.2.2  *Measuring actor engagement*

With the insight that actor engagement is multidimensional, including cognitive and affective/emotional dimensions and behavioural activities and manifestations that are formed by interactive processes dependent on the context and occurring over time, a complex concept emerges. Previous research on hospitality has measured online engagement. While there are deviating definitions of engagement, this is also reflected in how it is operationalized and measured. While So et al. (2014) identify five dimensions that commonly define the concept – enthusiasm, attention, absorption, interaction, and identification – Vivek et al. (2014) propose attention, enthusiastic participation, and social connection. In an early study on customer engagement, Patterson et al. (2006) emphasized vigour, dedication, absorption, and interaction as important dimensions. Mollen and Wilson (2010), studying online engagement with a brand, characterized engagement as captured through emotional congruence with the brand's online narrative together with the vigour of the cognitive processes in relation to utility and relevance. The emotional aspect was also part of Brodie et al. (2011) definition where passion was expressed by the respondents in relation to a brand. In combination with immersion (being absorbed in an activity; the individual is engrossed in something) and willingness to spend time and energy on an activity related to a brand (called activation), this reflects an individual's engagement in that brand. Previous and disparate ways of measuring engagement all have some commonalities, such that the cognitive aspect is reflected in awareness, knowledge, interest, being engrossed, or assimilation. On the affective side, the emotions evoked are reflected in measuring excitement or devotion. While the cognitive and affective aspects take place inside the mind and heart of the individual, the behavioural aspect shows through different activities such as willingness to recommend a place (WOM), to interact with others, or, as can be seen in a destination, in the form of exploring and staying longer in a place.

This can be summarized as in Table 12.1.

*Table 12.1* Classification of dimensions of engagements

| Engagement aspects | |
| --- | --- |
| Cognitive | Awareness |
| | Knowledge |
| | Interest |
| | Being engrossed |
| | Assimilation |
| Affective | Excitement |
| | Devotion |
| Behavioural | Willingness to recommend |
| | Exploration |
| | Length of stay |

## 12.3  Developing a method to measure multidimensional engagement

To capture the multidimensional concept of engagement in a destination/place, it is essential to understand the individual's different interactions and cognitive and affective processes in relation to where and when in a spatio-temporal dimension they take place. This calls for, as argued above, a broad approach in data collection to be able to capture the pre-, mid-, and end-of-visit journey of the individual.

### 12.3.1  *Designing data collection*

In understanding the whole process, two different datasets were used: survey and GPS loggers with a like-click function. In addition to these two, GSM data (mobile phone data) was used to better understand mobility changes at a meso-scale. While the latter category cannot be linked to survey or GPS responses, it does provide a generic depiction of visitor behaviours in the studied regions.

### 12.3.2  *Data collection in three different locations to test/verify the method*

Within the INCULTUM project, three different destinations in Sweden were selected for studying and testing the method for measuring engagement:

- Gotland. The largest island in the Baltic Sea, with the city Visby (read more at Toger et al., 2023b). Visby is characterized by its walled medieval city – a UNESCO World Heritage Site – which is on the verge of over-crowding during parts of the summer, but it has rural surroundings that could be developed further (Figure 12.1a).
- Öregrund – A picturesque coastal town in the region of Roslagen, rich in recreational amenities but not a "passer-by destination". It needs to develop its attractiveness to tourists (Figure 12.1b).

*Figure 12.1* Examples of the three studied sites: (a) Gotland (view from wall over the town); (b) The harbour area of Öregrund, Roslagen; (c) Drone photo of parts of the Torsö archipelago (west of Brommö).

- Torsö. An archipelago in the largest lake in the EU, a "stop-on-the-highway place" between Stockholm and Göteborg (Figure 12.1c) that needs to promote itself as an engaging destination to attract new potential inhabitants, as the population is decreasing.

### 12.3.3 Survey

As the purpose of this chapter is to understand engagement with a specific destination, the survey requires that the questionnaires are distributed to those who are visiting the area. A questionnaire can provide insights on pre-visit engagement, while also providing insight on the characteristics of the visitors. For each of the locations, we handed out a two-page, single-paper survey and pencils, for easy response in outdoor locations, to each party that agreed to participate.

Demographic information was collected to provide an understanding of who the visitor is and the reasons for the visit. To measure the cognitive aspect pre-visit, the questionnaire contained questions measuring previous knowledge about the destination, expectations, and any information search prior to the visit (referred to in Table 12.1 as Awareness, Knowledge, Interest, Being engrossed, and Assimilation). The questionnaire was also designed to measure mid-visit behaviour by asking the individual to state the places visited and the activities undertaken, such as whether the individual visited museums, specific sites, or a restaurant/café. A critique raised against questions asked post hoc is that they ask the individual to recall what they did during the visit (cf. Stewart & Hull, 1992). To overcome this, the volunteers were asked to take the questionnaire with them and fill it out during their visit. Another critique of questionnaires concerns using preformulated answer options, as alternatives important to the respondent may not be included. The questionnaire was therefore pre-tested and discussed with different stakeholders prior to data collection.

In measuring mid-visit behaviour, the unique opportunities and challenges that come with each of the regions make it difficult to develop identical methodological approaches for the collection of data. Therefore, slight alterations to the surveys were made to ensure that cross-location comparisons could be conducted at the same time as location-specific questions could be investigated. The main challenge in our study was, however, to select the best location and time for involving respondents. On Gotland, and in particular in Visby, tourist information desks, tourism information boards, web resources, etc., are widely available, but in Roslagen, and in particular on Torsö, there are few or no locations where all tourists would go for information gathering. This means that there are too many options on Gotland, and few or none at the other sites. In addition, finding one information locus implies that we only identify one type of visitor and not those who already have knowledge about the locations. To solve this, we mainly collected data in liminal spaces where visitors spend time before continuing their journey. For all of the selected sites, visitors wait for ferries (Torsö and Roslagen) or for ship-to-town transport (Visby). In addition to these locations, we also selected alternative data collection sites for sensitivity analysis.

To measure the end-of-visit behaviour and disposition to engage in the future, the questionnaire assesses satisfaction level (which is an estimate for Excitement in Table 12.1, an affective aspect), and willingness to recommend the destination (a behavioural aspect in Table 12.1).

### 12.3.4  GPS loggers

While much previous research measures engagement post hoc, there is still a lack of research measuring engagement mid-visit in real time. The inclusion of this analytical dimension is essential as the experience of an individual visiting a place develops and is affected by the processual spatio-temporal dimension (Aho, 2001; Caldeira & Kastenholz, 2019; Godovykh & Tasci, 2020). Recent research has introduced methods for tracking tourists' movements with new technology such as GPS loggers (De Cantis et al., 2016; Ferrante et al., 2018). While the technology makes it possible to follow in the tourists' footsteps, this technology has primarily been used to understand the consumption of place, i.e. the movements per se. Although adding to the knowledge on how individuals move in a place, how long they stay and where they go, use of the technology for understanding the engagement process is new. The GPS loggers used in the INCULTUM project are the size of a USB flash drive and have a button on the top that can be used for different purposes, in this case to capture what the individuals like (a "like-click" button) – see Figure 12.2. This technology offers the opportunity to capture how tourists move and interact with the place, how long they stay at different points of interest, and, when given the instruction, it shows what they like as they click on the button (Figure 12.2c).

We asked a member of each party to wear a GPS logger as a necklace and instructed the volunteers to like-click locations that the travelling party enjoyed (see details below). Since each GPS logger has a unique number noted in a designated part of the survey, trajectories and survey responses could be connected. We selected times according to ferry/ship arrival timelines and made sure to be available for collection of GPS loggers and survey responses upon the participants' return. We also selected days with similar weather, to make sure that activities and choices were comparable between locations and over time.

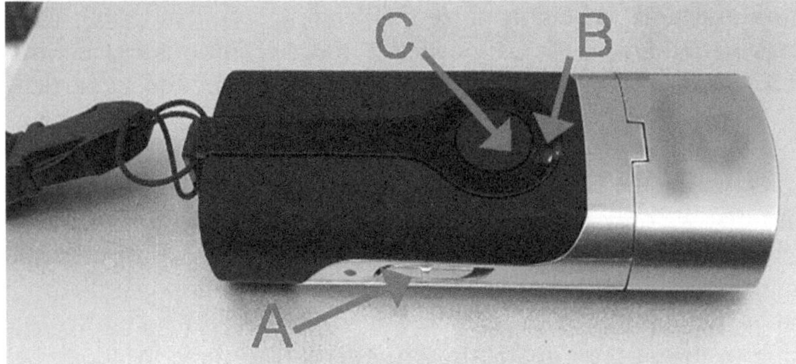


*Figure 12.2* GPS logger.

*Table 12.2* Classification of dimensions of engagement and temporal subdivisions of the tourist experience

| Construct dimensions | Type of data | Pre-visit | Mid-visit | End of visit |
|---|---|---|---|---|
| Cognitive | Survey | x | | |
| | GPS | | x | |
| | Survey | | | x |
| Affective | GPS like-click | | x | |
| | Survey | x | | |
| | Survey | | | x |
| Behavioural | GPS | | x | |
| | GSM | | x | |

To measure mid-visit engagement, the GPS loggers' like-click function is used as an estimate for affective and behavioural aspects of engagement. While like-clicking on a general level can work as an indicator of engagement, it is hard to estimate to what extent the level is high or low, unless also considering the number of like-clicks in relation to attention and vigour. The reason for this relative measure is that an individual's like-clicking once during a 7-hour visit could be argued to indicate relatively less engagement in the destination than an individual like-clicking once during a 1-hour visit. Attention is here operationalized as duration, in this case in the form of time spent at the destination, whereas vigour is measured as distance travelled during the visit.

In Table 12.2, the multidimensional classifications of engagement, as suggested in the literature section, are situated from pre-, mid-, and end-of-visit perspectives, and the method for data collection presented for each perspective. The table output suggests that while surveys are central, the combination of surveys and GPS loggers is essential for the understanding of engagement. Preliminary studies suggest (under review) that there is a significant positive correlation between the number of like-clicks and the stated value indicating how pleased the respondents were with their visit.

In addition to the relationship between surveys and GPS logging, it is also possible to draw conclusions from survey responses and the observed GSM behaviour over time, although the temporal dimensions became very different. In Table 12.3, responses from the 2021 survey from the Torsö archipelago revealed long-term behavioural changes in preferences as a result of the pandemic. GSM results confirmed the survey results (see Figure 12.5).

Using a combination of GPS trajectories and questionnaires, with the latter's answers matched to amenities and features identified in GIS resources (here collected through the GPS logger), we can match spatial activity to engagement. Starting with the GPS material, the GPS trackers

*Table 12.3* Survey responses from Torsö 2021 to the question about how and when they have changed their tourism/recreation behaviour because of the COVID-19 pandemic

| *How did the pandemic affect your behaviour?* | *Last year (%)* | *This year (%)* | *Next year (%)* | *Not at all (%)* |
| --- | --- | --- | --- | --- |
| Avoiding crowds | 36 | 44 | 9 | 16 |
| Avoiding cities | 14 | 13 | 5 | 6 |
| Using private vehicle | 19 | 25 | 6 | 7 |
| Outdoor activities preferred | 26 | 33 | 6 | 5 |
| Nature important | 19 | 27 | 5 | 6 |

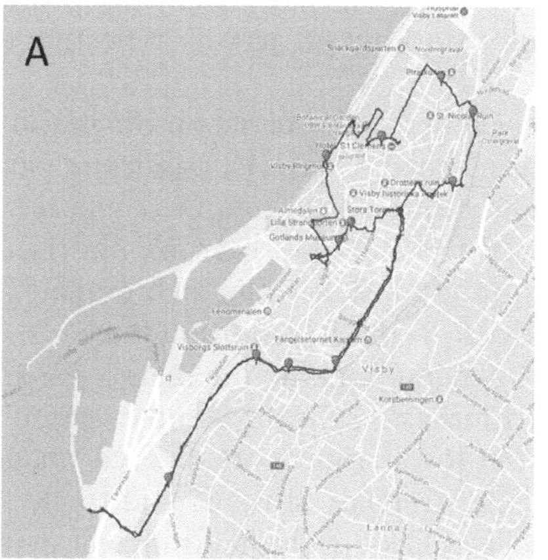
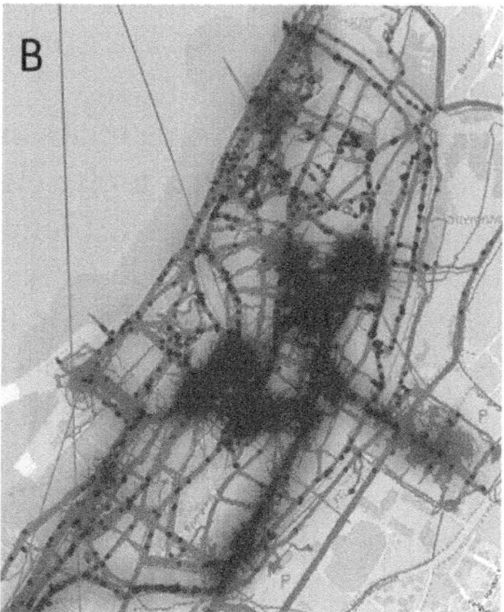

*Figure 12.3* GPS trajectories and like-clicks in Visby (Gotland).

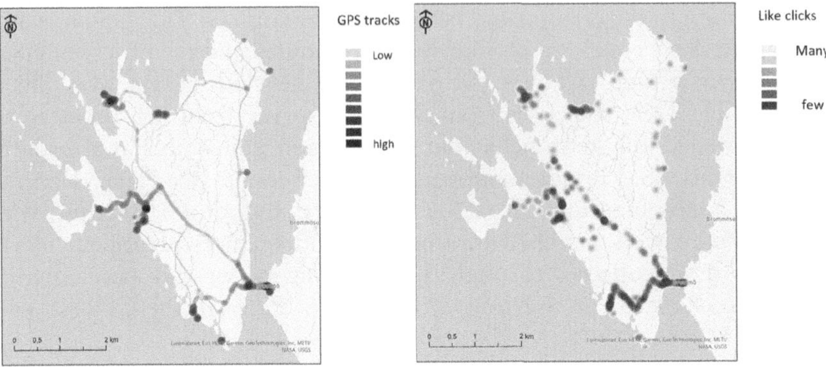


*Figure 12.4* GPS trajectories and like-clicks on the island of Brommö in the Torsö
archipelago.

render results as shown in Figure 12.3a, where the line shows the trajectory chosen by a randomly selected tourist to Visby, Gotland, and the
needle points indicate locations where the tourist has like-clicked. In Figure 12.3b, a full set of trajectories from tourists to Visby are overlaid. The
patterns reveal two things about the visitors' behaviour: (1) the visitors
are more or less keeping to the town streets of Visby, and do not wander
beyond the medieval town wall that encircles the core part of Visby, and
(2) the spatial distribution of GPS tracks and like-clicks are similar, suggesting that like-clicking is an activity conducted in the moment, and not
an activity *saved to a later or better moment*, but an activity reflecting the
experience of the moment.

In Figure 12.4, trajectories and like-clicks from tourists on the island of
Brommö in the Torsö archipelago during the COVID-19 pandemic (summer of 2021) are shown. Using a slightly different technique compared to
that in Figure 12.3, the density of trajectories and like-clicks are illustrated.
Brommö Island is known for its long sandy beaches and rich fauna. The
few private motorized vehicles that exist on the island are allowed to be
used only by islanders (and property owners), making the mainly car-free
island a cycling and hiking haven for tourists. From the common point of
entry (the ferry connection, seen in the lower right part of each image),
the visitor trajectories strictly follow the stretches of roads and tracks on
the island (left-hand image), and the like-clicks (right-hand image) depict
a similar pattern. The greatest density of tracks can be seen leading to the
four main beach areas on the island (two in the north, one to the west, and
one to the south). The patterns reflect what we observed in Figure 12.3, i.e.
the visitors follow the designated paths, and like-click scenic locations as
they travel.

### 12.3.5   *GSM mobility patterns*

Uppsala University hosts the GSM database MIND, which contains pseudonymized, longitudinal records of GSM mobility. The database allows us to trace phone mobility patterns over limited time spans, but the spatial detail is limited to the location of the GSM antennas, which means that we can associate phones with areas and also match these areas to contextual statistics or land use data from GIS resources, but we cannot match locations of phones to addresses or specific amenities. By aggregating phones per spatial unit, and by comparing the relative density over time, we can determine whether individuals (phone users) are more or less likely to spend time in areas of different recreational qualities during vacation times, regular weekdays, and weekends. The association between phones and amenities was elucidated in several steps: (1) GSM antenna locations were transformed to km² units in which the phones were located; (2) mapped land use features from OpenStreetMap were used to calculate the dominance (area dominance using focal statistics in GIS) of different kinds of green and blue amenities in each of the GSM-designated km² units; (3) study preferences in land use exposure on different days, or as a change in preferences over time. In a series of papers, the GSM material described has

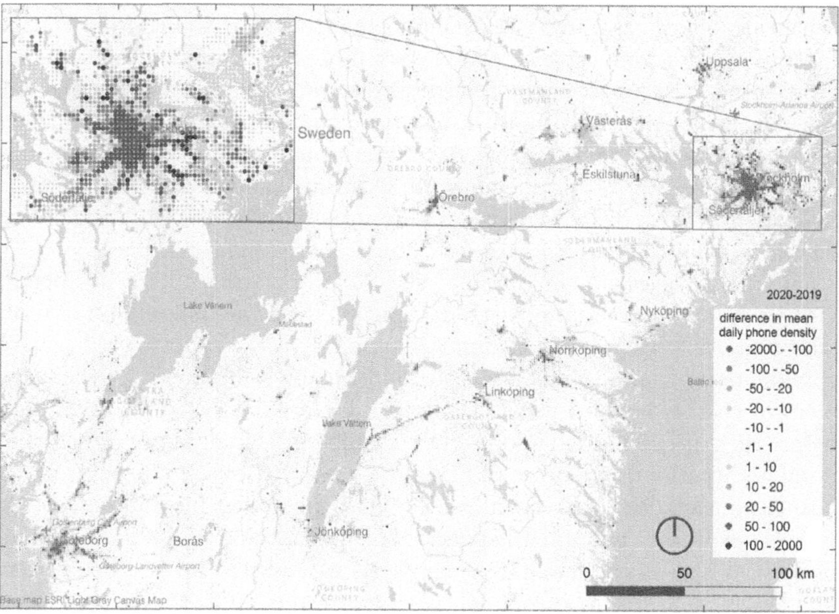


*Figure 12.5* Difference in population density between July 2020 and July 2019, in the central parts of Sweden. Darker colours indicate greater population change (for a full colour version of map cf. Östh et al., 2023). Map produced with ESRI ArcGIS®, base map ESRI Light Gray, map data sources: Lantmäteriet, Esri, Garmin, GeoTechnologies, Inc., METI/NASA, USGS.

been used to depict recreational mobility behaviour in Sweden. In Östh et al. (2023), the changing choice of locations during the pandemic is studied. They used a first differencing approach to show the relative change in population density during a Thursday in July, the main vacation period in Sweden (see Figure 12.5). The figure clearly shows how individuals are avoiding densely populated regions and favouring remote areas, and especially areas close to water, during the pandemic. The selection of nature amenities studied by Östh et al. (2023) was based on the survey conducted on Torsö.

Toger et al. (2023a) and Shuttleworth et al. (2023) find similar patterns where rural amenities are favoured during weekends, and during holidays such as Easter. Both of these studies include the Roslagen archipelago in the Baltic Sea, and both studies concluded that the archipelago gained population during recreational periods. The results are not spatially detailed enough to help us understand the engagement of individual visitors, but it helps us to understand the general change of behaviour in broad brushstrokes.

## 12.4    Concluding discussion

A cornerstone of our chapter is to reason about the nature of visitors' engagement and how to design the methodological setup for the measurement of their engagement in rural settings. We exemplify our discussion about visitor engagement using data from three rural and coastal but otherwise relatively different locations in Sweden. For each of the three regions, we have employed the same methods, which primarily means using survey and GPS logging techniques for the measurement of tourist engagement, but by adding GSM data, we are also able to show how, over time, changes in locational patterns for phones can be matched to survey responses, and where proximity to nature, avoidance of densely populated areas, etc., can be traced in both kinds of datasets.

What can we learn from our analysis? We propose that the engagement of visitors at tourist destinations is best measured using a combination of tools, where the different temporal dimensions of a visit are considered using a combination of survey and GPS measurements. By using surveys, the engagement of visitors before (when initially handed the survey, if this takes place before the trip) and after (upon return) can be measured. However, the temporal and spatial distribution of like-clicks observed by GPS loggers and the positive correlation between like-clicks and the stated pleasantness of the visit (as expressed in surveys) suggest that like-clicking on GPS devices and using GPS loggers can facilitate the analysis of engagement, and is suggested as a key method for measuring mid-visit engagement. On an aggregate level, the results from the GPS studies also reveal another important result. There seems to be a strong link between the duration of stay and like-clicking, in that areas with more and longer stays have a corresponding increase in engagement. Although the causal relationship between stay and like-click can be difficult to tease out using our technique, the results suggest that unattractive areas are not only not "liked", but they are also quickly passed.

### 12.4.1   Limitations

An issue with our studies of engagement is that data collection took place during different phases of the pandemic. The pandemic affected peoples' behaviour to a considerable extent, and in ways that challenged behavioural norms and distribution of visitors. The pandemic enabled us to study the change in behaviour triggered by a shock, and as such, our studies were rewarding in providing evidence of a change in preferences towards more nature and solitude at times of risk. However, we have been unable to specify a state of *normal* behaviour during our study period. Uncorrelated, but worth noting, are also the effects of climate change, which is rapidly affecting tourism. Over a five-year period (2018–2023), Torsö experienced the warmest summer and the rainiest vacation period in a 100-year period, and the other study regions were almost as heavily hit. This underlines the challenge of finding a normal behaviour among tourists, and it also highlights a long-term challenge for tourism development.

### 12.4.2   Suggestions for future studies

The method was developed and tested in three locations in Sweden. As behaviours are context-specific, testing in other contexts is important for refining the method. In this study, the pandemic may have affected the results. Future studies would also benefit from an analysis of the resilience of destinations, in terms of how well they are coping with or mitigating the effects of both the pandemic and increasing climate changes, and how these are affecting visitor behaviour.

### Note

Maps throughout this chapter were created using ArcGIS® Pro software by Esri. ArcGIS® is the intellectual property of Esri and used herein under license. Copyright © Esri. All rights reserved. For more information about Esri® software, please visit www.esri.com.

### References

Aho, S. K. (2001). Towards a general theory of touristic experiences: Modelling experience process in tourism. *Tourism Review*, 56(3/4), 33–37.

Bergel, M., & Brock, C. (2019). Visitors' loyalty and price perceptions: The role of customer engagement. *The Service Industries Journal*, 39(7–8), 575–589.

Brodie, R. J., Fehrer, J. A., Jaakkola, E., & Conduit, J. (2019). Actor engagement in networks: Defining the conceptual Domain. *Journal of Service Research*, 22(2), 173–188.

Brodie, R. J., Hollebeek, L. D., Jurić, B., & Ilić, A. (2011). Customer engagement: Conceptual Domain, fundamental propositions, and implications for research. *Journal of Service Research*, 14(3), 252–271.

Caldeira, A. M., & Kastenholz, E. (2019). Spatiotemporal tourist behaviour in urban destinations: A framework of analysis. *Tourism Geographies*, 22(1), 22–50.

Chandler, J. D., & Lusch, R. F. (2015). Service systems: A broadened framework and research agenda on value propositions, engagement, and service experience. *Journal of Service Research*, 18(1), 6–22.

Cöster, M., Gebert-Persson, S., & Ronström, O. (2023). *Enabling Sustainable Visits*. Visby: Eddys förlag.

De Cantis, S., Ferrante, M., Kahani, A., & Shoval, N. (2016). Cruise passengers' behavior at the destination: Investigation using GPS technology. *Tourism Management*, 52, 133–150.

Dessart, L., Veloutsou, C., & Morgan-Thomas, A. (2016 March 23). Capturing consumer engagement: duality, dimensionality and measurement. *Journal of Marketing Management*, 32(5–6), 399–426.

Fehrer, J. A., Woratschek, H., Germelmann, C. C., & Brodie, R. J. (2018). Dynamics and drivers of customer engagement: within the dyad and beyond. *Journal of Service Management*, 29(3), 443–467.

Ferrante, M., Magno, G. L. L., & De Cantis, S. (2018). Measuring tourism seasonality across European countries. *Tourism Management*, 68, 220–235.

Frechtling, D. C. (2006). An assessment of visitor expenditure methods and models. *Journal of Travel Research*, 45(1), 26–35.

Godovykh, M., & Tasci, A. D. (2020). Customer experience in tourism: A review of definitions, components, and measurements. *Tourism Management Perspectives*, 35, 100694.

Hamilton, R., Ferraro, R., Haws, K. L., & Mukhopadhyay, A. (2021). Traveling with companions: The social customer journey. *Journal of Marketing*, 85(1), 68–92.

Harmeling, C. M., Moffett, J. W., Arnold, M. J., & Carlson, B. D. (2017). Toward a theory of customer engagement marketing. *Journal of the Academy of Marketing Science*, 45, 312–335.

Harrigan, P., Evers, U., Miles, M., & Daly, T. (2017 April 01). Customer engagement with tourism social media brands. *Tourism Management*, 59, 597–609.

Hollebeek, L. D., Srivastava, R. K., & Chen, T. (2019). S-D logic–informed customer engagement: integrative framework, revised fundamental propositions, and application to CRM. *Journal of the Academy of Marketing Science*, 47(1), 161–185.

Kumar, V., Aksoy, L., Donkers, B., Venkatesan, R., Wiesel, T., & Tillmanns, S. (2010). Undervalued or overvalued customers: Capturing total customer engagement value. *Journal of Service Research*, 13(3), 297–310.

Kumar, V., & Pansari, A. (2016). Competitive advantage through engagement. *Journal of Marketing Research*, 53(4), 497–514.

Li, L. P., Juric, B., & Brodie, R. J. (2018). Actor engagement valence: Conceptual foundations, propositions and research directions. *Journal of Service Management*, 29(3), 491–516.

Mollen, A., & Wilson, H. (2010). Engagement, telepresence and interactivity in online consumer experience: Reconciling scholastic and managerial perspectives. *Journal of Business Research*, 63(9), 919–925.

Östh, J., Toger, M., Türk, U., Kourtit, K., & Nijkamp, P. (2023). Leisure mobility changes during the COVID-19 pandemic – An analysis of survey and mobile phone data in Sweden. *Research in Transportation Business & Management*, 48, 100952.

Patterson, P., Yu, T., & De Ruyter, K. (2006 December). Understanding customer engagement in services. In Advancing theory, maintaining relevance. Proceedings of ANZMAC 2006 conference, Brisbane (pp. 4–6).

Quaranta, G., Citro, E., & Salvia, R. (2016). Economic and social sustainable synergies to promote innovations in rural tourism and local development. *Sustainability*, 8(7), 668.

Rather, R. A., Ramkissoon, H., Hollebeek, L. D., & Loureiro, S. M. C. (2023). Customer engagement in tourism and hospitality research. In Rather, R. A. & Ramkissoon, H. (eds.), *Handbook of Customer Engagement in Tourism Marketing* (pp. 1–24). Cheltenham: Edward Elgar Publishing.

Shuttleworth, I., Toger, M., Türk, U., & Östh, J. (2023). Did liberal lockdown policies change spatial behaviour in Sweden? Mapping daily mobilities in Stockholm using mobile phone data during COVID-19. *Applied Spatial Analysis and Policy*, 17, 1–25.

Smith, S. (2000). New developments in measuring tourism as an area of economic activity. In W. C. Gartner and D. W. Lime (eds). *Trends in Outdoor Recreation, Leisure and Tourism* (pp. 225–234). CAB International.

So, K. K. F., King, C., & Sparks, B. (2014). Customer engagement with tourism brands: Scale development and validation. *Journal of Hospitality & Tourism Research*, 38(3), 304–329.

Stewart, W. P., & Hull IV, R. B. (1992). Satisfaction of what? Post hoc versus real-time construct validity. *Leisure Sciences*, 14(3), 195–209.

Storbacka, K. (2019). Actor engagement, value creation and market innovation. *Industrial Marketing Management*, 80, 4–10.

Storbacka, K., Brodie, R. J., Böhmann, T., Maglio, P. P., & Nenonen, S. (2016). Actor engagement as a microfoundation for value co-creation. *Journal of Business Research*, 69(8), 3008–3017.

Toger, M., Türk, U., Östh, J., Kourtit, K., & Nijkamp, P. (2023a). Inequality in leisure mobility: An analysis of activity space segregation spectra in the Stockholm conurbation. *Journal of Transport Geography*, 111, 103638.

Toger, M., Östh, J., & Persson, S. G. (2023b). What you see is where you go: Cruise tourists' spatial consumption of destination amenities. *Economic Themes*, 61(1), 63–84.

UNWTO. (2023). *Glossary of Tourism Terms*. Retrieved from https://www.unwto.org/glossary-tourism-terms, 230523.

van Doorn, J., Lemon, K. N., Mittal, V., Nass, S., Pick, D., Pirner, P., & Verhoef, P. C. (2010). Customer engagement behavior: Theoretical foundations and research directions. *Journal of Service Research*, 13(3), 253–266.

Vivek, S. D., Beatty, S. E., Dalela, V., & Morgan, R. M. (2014). A generalized multidimensional scale for measuring customer engagement. *Journal of Marketing Theory and Practice*, 22(4), 401–420.

Zhang, H., Gordon, S., Buhalis, D., & Ding, X. (2018). Experience value cocreation on destination online platforms. *Journal of Travel Research*, 57(8), 1093–1107.

# 13 Cultural tourism, cutting-edge technologies and participatory planning in the context of worth-living integrated development

*Vaios Kotsios and Sotiris Tsoukarelis*

## 13.1 Introduction

The INCULTUM project explores the challenges and opportunities in the field of cultural tourism, with the aim of enhancing social, cultural and economic development, exploiting the real potential of remote areas, by the local communities and anyone who is interested. In particular, nine methodological approaches are developed, with active participation of all stakeholders at the core, for the development of cultural tourism in areas with different geographical, socio-economic and cultural characteristics.

This chapter aims at presenting the innovative approaches of Integrated Outputs, used for the recording, analysis, synthesis and visualization of the natural and socio-economic reality of the region, aiming at the development of cultural tourism under the prism of Worth-living Integrated Development (Rokos, 2004a, b), utilizing cutting-edge technologies and the integrated infrastructure system and policies for environment and development IDPSS (Kotsios, 2016).

The methodological approach consists of the following steps:

1 Creating a knowledge geobase for the region, linked to business intelligence tools, with information on the set of data on its physical and socio-economic reality.
2 Conducting empirical research to collect primary data on the constraints, problems, opportunities and prospects for the development of cultural tourism in the region.
3 Entering the data into the integrated cadastral infrastructure system for the environment and development (IDPSS) and obtaining actions for the development of cultural tourism in the study area.

## 13.2 Cognitive Geobase

### 13.2.1 Secondary-administrative data

For the analysis of the natural and cultural resources, data were collected, with a brief description and the relevant bibliographic reference where the

DOI: 10.4324/9781003422952-13

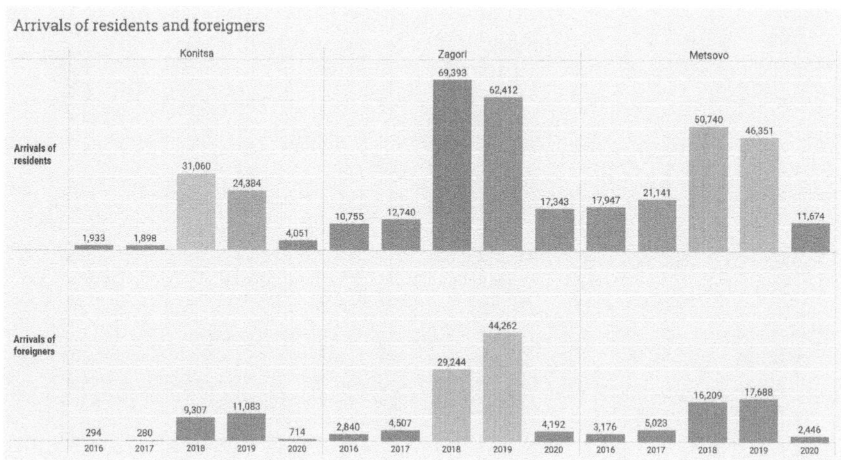


*Figure 13.1* Arrivals of residents and foreigners.
*Source:* ELSTAT

information comes from and the coordinates of each resource. In addition, accompanying images were collected for each of these elements so that users could view the images in combination with the interactive map of the sites.

In order to achieve a proper use of the data, we began with the necessary corrections, the grouping of the data into specific columns in the database and finally we carried out a "join" of the tables in order to synthesize the information. In addition, for most of the data, the necessary links for the Google Street View were collected. Where possible through Google Maps, the location was identified and the relevant link was recorded and, as a result, users of the interactive map have access to the street view.[1]

For the analysis of the economic reality of the study area in the field of cultural tourism, data from ELSTAT and the internet (Google Trends) were used in a comparative analysis of three regions with common characteristics (Konitsa, Metsovo and Zagori). As can be seen in Figure 13.1, foreign and domestic **arrivals** over time are mainly related to the Municipality of Zagori.

In addition, **online searches** for the area were investigated. Also, Figure 13.2 shows some connecting diagrams indicating the total searches per month over the last five years based on four specific keywords in Greek and Latin characters. The study area showed variation in terms of the searches in December and August and the search language. As shown in the chart below, searches for the term "Konitsa" in Latin characters rank much lower in searches.

At the **working level,** a comparative analysis of the structure of salaried employment in business activity in three regions (Konitsa, Zagori and Metsovo) was carried out. It was observed that the structure of entrepreneurship and salaried employment in the region of Konitsa is distributed across other

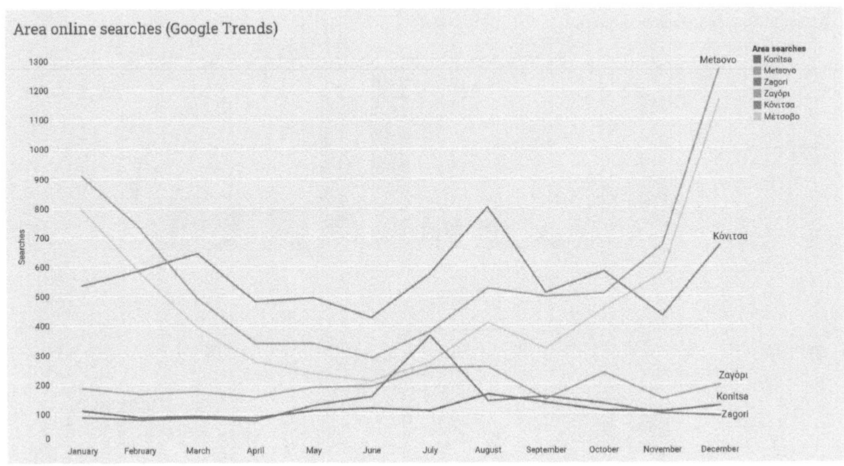


*Figure 13.2* Area online searches.

*Source:* Google Trends

sectors and is not highly concentrated in the accommodation and food services sector as in other neighbouring destinations with similarly high cultural and environmental reserve.

At the same time in Konitsa, one-fourth of employment is in wholesale and retail trade and one-fifth in human health and social work, while these activities in the neighbouring destinations are largely absent. In Konitsa, 464 people are in the registry of whom 246 are women and 218 are men. It appears that the unemployed are found across the age grid with a higher concentration in the 40–49 age group. Half of them are young unemployed. We also note that of the registered unemployed, half are unmarried and the vast majority have no children.

As documented later, based on the research conducted among the Cultural Associations of the Municipality of Konitsa, as well as from previous research of the ME.K.D.E., it seems that demographic issue is a major and common problem for all mountainous regions.

## 13.3 Empirical research to collect primary data on the constraints, problems, opportunities and prospects for the development of cultural tourism in the region

Along with the collection of secondary data, qualitative empirical research was conducted using structured interviews with the cultural associations of the region (see Annex), in order to collect critical data for the development of cultural tourism in the region. Research results were presented and discussed with the associations themselves, in order to take specific actions, in the context of the participatory process of promoting collective intelligence of the pilot project "Aoos, the shared river".

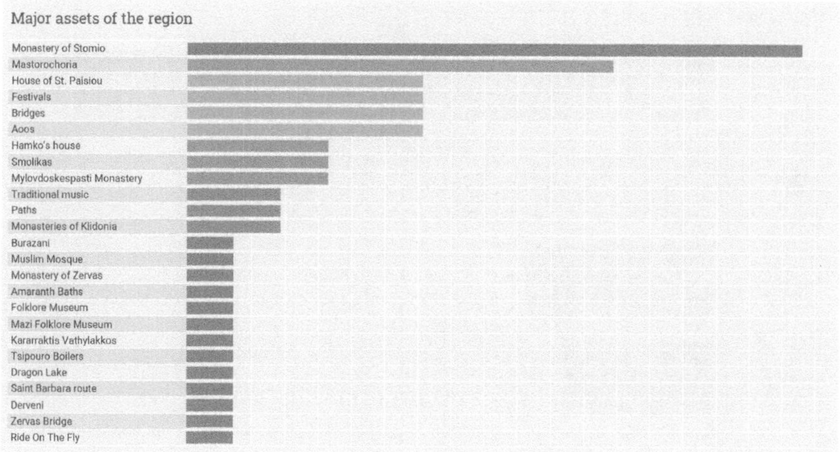


*Figure 13.3* Major assets of the region.

On the question "what in your opinion are the most important cultural and environmental assets of the region?", it was found that the main assets were the Monastery of Stomio, which was mentioned by almost all participants, Mastorochoria, i.e. all the villages that make up the old Municipality of Mastorochoria, the house of Agios Paisios, Panigiria, as a cultural production activity, the bridges, the house of Hamko, Smolikas, and the monastery of Molybdsvodskepastis. Other important assets mentioned by the Associations are traditional music, the trails, the monastery of Klidonia, Bourazani, the Muslim mosque, and the monastery of Zermas, the Amarantou baths, the folklore museums, the waterfalls in Vathylakkos, the cauldrons as a cultural activity, the Drakolimnes, the route of Agia Varvara, Derveni, the Zermas bridge (Figure 13.3).

Asked to highlight the most important problems of the region, the Associations indicated infrastructure as a major problem. The road network and the health and education infrastructure in the region appeared to be problematic. A second major issue raised by the Associations' responses is the issue of demographic depopulation, which is directly linked to employment and unemployment in the region. A third issue also linked to these is social activity. The fourth issue in order is unemployment (Figure 13.4).

There are many unused infrastructures found in the region that could be reused. According to the responses to the survey, the most important one available for use is the Anagnostopouleio Agricultural School. The second unused resource mentioned is the old leather and tannery factory, followed by schools that exist in almost every village, the cultural centre in Plagia and the café that has rooms for the operation of a guest house, the traditional bridge in Elefthero. Also an unused infrastructure seems to be the paths and festivals without microphone "in the plain", as mentioned by the Associations (Figure 13.5).

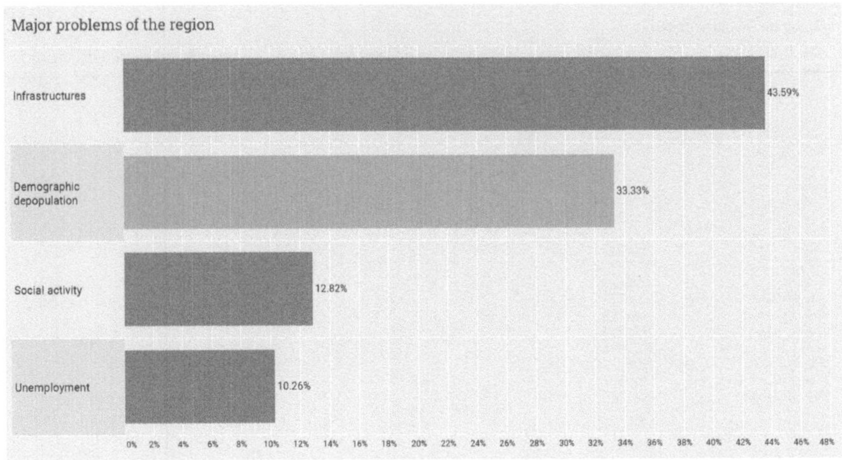


*Figure 13.4* Major problems of the region.

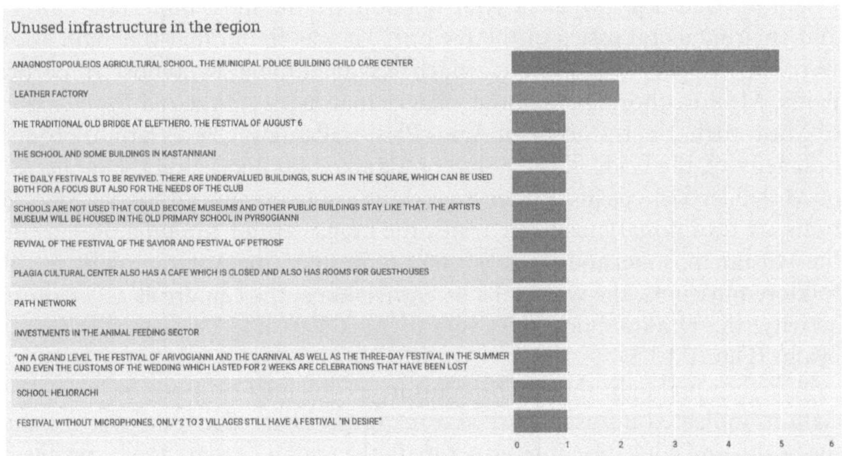


*Figure 13.5* Unused infrastructure in the region.

When asked which products can be considered as traditional, quality products in the region, the first one mentioned was melon. The second product was cheese. However, in the district of Konitsa, there is not a single cheese dairy or cottage industry. Other products such as peaches, giant legumes, aromatic plants, sea buckthorn, fodder, tsipouro, cattle, trahanas, pies, trout, apples, honey, jams, vegetables and nuts follow (Figure 13.6).

Although one of the major problems highlighted by the survey is demographic decline, according to the responses of the cultural associations, there are possibilities for reoccupation (62%). One in four was not certain and only 12% said there are no such possibilities in the area.

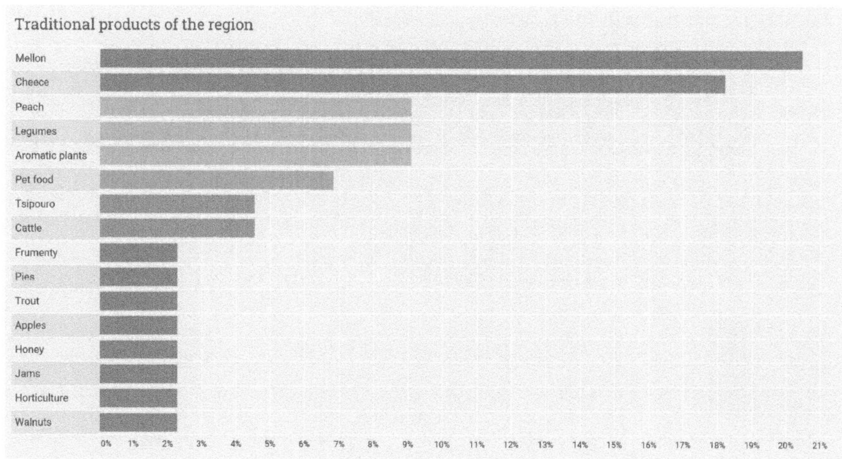


*Figure 13.6* Traditional products of the region.

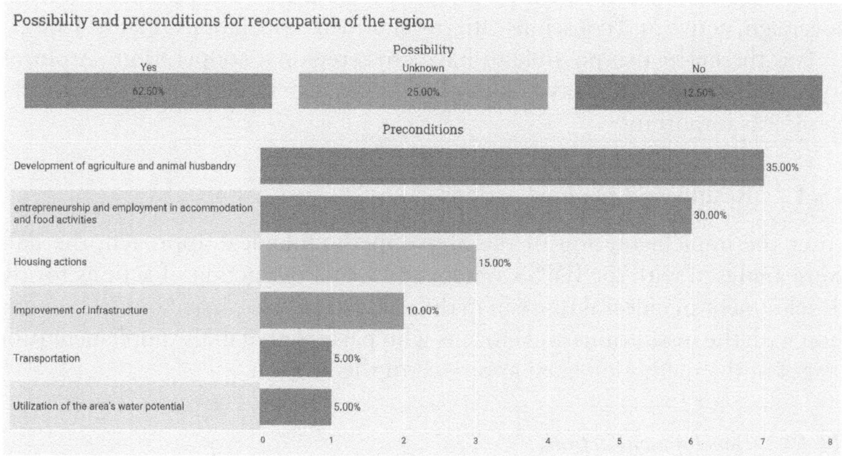


*Figure 13.7* Possibility and preconditions for reoccupation of the region.

The preconditions for reoccupation, according to the survey, are the development of agriculture and livestock, entrepreneurship and employment in accommodation and food services sector, which is very low, the utilization of housing, improved infrastructure and transport. Finally, another prospect mentioned by the Associations is the exploitation of the aquatic potential of the region (Figure 13.7).

When asked if the Associations could cooperate with each other, two out of three Associations were positive, which is very encouraging, only one out of three said that there can be no cooperation between the Associations. An important parameter: in order to create another reality for the region, at the core

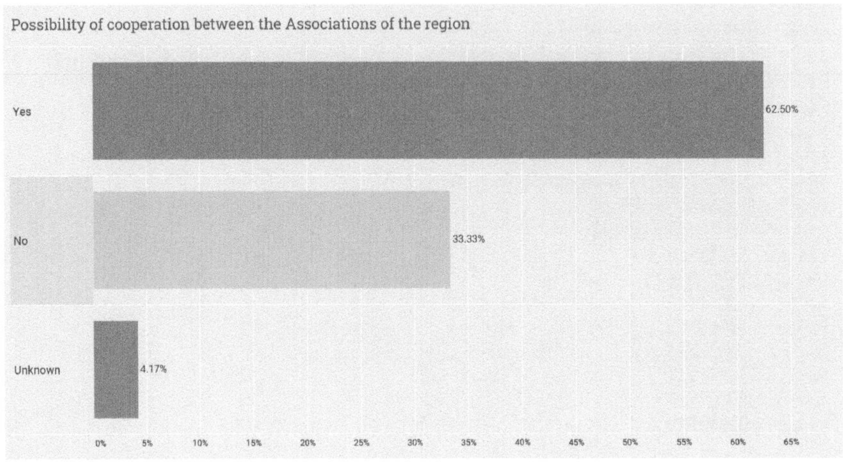


*Figure 13.8* Possibility of cooperation between the Associations of the region.

of an integrated development philosophy, there must be cooperation between sovereign, active and conscious citizens and their associations (Figure 13.8).

Is it therefore also possible to have cross-regional cooperation? A plurality, less than half, 45% say it is difficult in practice to do so. But some, 40%, think it is important.

## 13.4    Results and planned actions

After the implementation of the above methodological approach, the data were analysed with the IDPSS system and a first extraction of actions for the development of cultural tourism in the study area was carried out in cooperation with the institutions and citizens who participated in the implementation stages of the methodological approach of the project:

### 13.4.1    Development actions

- Programmatic Agreement with the Municipality of Konitsa (associated partner of the project) for the exploitation of the Mansion of Hamko (unused resource according to the research), through the unique collection of rare records of traditional music of the region and the Balkans (unused resource according to the research), of the Grammy award-winning musicologist and collector Christopher King who resides in Konitsa.
- Development of an Open Source Water Monitoring Tool (waterjet[2]) in collaboration with the associated partners of the project P2PLab and researchers from New Dexterity, in order to protect the local water resources and boost scientific tourism in the area. Taking advantage of the "design global, manufacture local" model, which builds on the conjunction of the digital commons of knowledge and design with desktop and

benchtop manufacturing technologies (from three-dimensional printers and laser cutters to low-tech tools and crafts), we decided the organization of three workshops using the facilities and equipment of the public library of Konitsa in order to locally produce a water monitoring tool called "waterjet". The local community was involved in the 3D printing process of the waterjet and trained on the use and the maintenance of the tool. The waterjet is available to scientists who want to conduct research on the water resources of the area and to the local society in order to monitor the condition of their water environments. Aoos River and the Dragonlake of Tymfi where identified as major natural assets according to the research.

- Opening and recording of a hiking path[3] leading from the village of Kallithea to the Agios Konstantinos chapel (cultural heritage monument), identified by the Cultural Association of Kallithea. The action will be implemented by using network tools such as "work away" as a way of addressing problems highlighted by the research, such as lack of human and financial resources through the use of solidarity and voluntary tourism.
- Recording of a Transregional pastoral route[4] in cooperation with the other pilot of the project "Vjosa, the shared river" in the framework of the cross-border piloting and in the joint promotion of Vlach culture, pastoralism and transhumance.
- Inventory of licensed distilleries in the area (cultural asset according to the research), their owners, their capacity and their supporting facilities for the organization of a relevant cultural tourism product.
- Collaboration with social economy actors in the region to create a web platform[5] for the promotion of the region (searches for the term "Konitsa" in Latin characters rank much lower than neighbouring areas in Google Trends according to the research), as well as a participatory design of cultural experiences, aimed at creating supply, demand and evaluation of cultural products and services.

## Appendix: cultural associations that participated in the survey

- Rural Cultural Association of Aetopetra
- The Brotherhood of Agia Paraskevi of Konitsa "Kerasovo"
- The Kerasovo Brotherhood of Kiato "Zarkadi"
- Konitsa Sports Association
- Social Cooperative Enterprise Kaleidoscope
- Cultural-Mountaineering-Improvement and Sports Association of Paleoselli
- Improvement and Progressive Association of Konitsa
- Improvement-Cultural Association of Mazi "Aoos"
- Educational-Cultural Association of Heliorachi
- Cultural and Improvement Association of Elefthero Konitsas.
- Museum of Ali Pasha and Revolutionary Era

- Mountaineering club
- Cultural-Improvement-Folklore Association of Zerma (Plagia) "The passing of Theotokos"
- Cultural Association of Amarantos
- Cultural Association of the Vlachs of distrato
- Cultural Association of Vourbiani
- Cultural Association of Kalithea of Konitsa " Goritsa"
- Cultural Association of young people of Konitsa
- Cultural-Sports Association of Kledonia "Litoviani"
- Cultural Developmental Improvement Association of Gorgopotamos
- Progressive Union Pyrsoyanni
- Progressive Charitable Association of Kastaniani "Agios Nicolaos"
- Women's Association
- Association of The Fraternity of Oxias of Epirus "Agios Nicolaos"
- Dance Association of Konitsa

## Notes

1 The interactive map "Aoos, the Shared River" is available online: https://public. tableau.com/app/profile/social.analytics.gr/viz/Aoosthesharedriver_RS_Update/ Aoosthesharedriver.
2 https://newdexterity.org/autoboat/.
3 https://www.wikiloc.com/hiking-trails/cultural-route-in-kallithea-by-the-high-mountains-incultum-116151825.
4 https://www.wikiloc.com/hiking-trails/vlachs-transhumance-path-grammos-by-the-high-mountains-incultum-140975712.
5 https://culturaltourism.gr/.

## Bibliography

INnovative CULtural ToUrisM in European peripheries (INCULTUM, 2022). https://incultum.eu/

Kostakis, V., Latoufis, K., Liarokapis, M., & Bauwens, M. (2018). The convergence of digital commons with local manufacturing from a degrowth perspective: Two illustrative cases. *Journal of Cleaner Production*, 197, Part 2, 1684–1693.

Kostakis, V., Niaros, V., & Giotitsas, C. (2023). Beyond global versus local: Illuminating a cosmolocal framework for convivial technology development. *Sustainability Science*, 18, 2309–2322. https://doi.org/10.1007/s11625-023-01378-1

Kotsios, V. (2016). *Creating an Integrated System of Cadastral Infrastructure, Environment and Development*. PhD Thesis, National Technical University of Athens (NTUA).

Rokos, D. (2004a). *The Integrated Development of Epirus. Problems, Possibilities and Constraints*. 4th Interdisciplinary Interuniversity Conference of the H.M.P. and the ME.K.D.E. of the H.M.P. The Integrated Development of Epirus Metsovo, 23–26 September 2004.

Rokos, D. (2004b). *Integrated Development in Mountainous Areas*. Theory and Practice, in Proceedings of the 3rd Interdisciplinary Interuniversity Conference of the H.M.P. and the H.M.P.'s ME.K.D.E. "Integrated Development in Mountainous Areas. Theory and Practice", Metsovo Conference Centre, 7–10 June 2001, pp. 79–140.

# 14 Experimental reconstruction, cultural memory and cultural tourism

## Vlach minority heritage in the Upper Vjosa Valley, Southern Albania

*Eglantina Serjani and Ardit Miti*

## 14.1 Introduction

In the 2010s, while undertaking archaeological field research in the Upper Vjosa Valley, southeastern Albania, we initiated our interest in the Latin-speaking community of the Vlach. Historically, they inhabited swathes of the Balkans but are predominantly located within southern and central-western Albania, north-west and central-eastern Greece, and western and central North Macedonia. The name Vlach is assumed to derive from Slavic as a designation for Latin speakers, originating from the Germanic word *Wälsche* meaning 'stranger' (Dvoichenko-Markov, 1984, p. 525; Winnifrith, 1987, pp. 1–35). Other names have been given to them by locals, such as *Çoban* in Albania, *Kutsóvlachi* and *Arvanitóvlachi* in Greece, *Macedo-Romanian* in Romania or *Tsintsar* and *Vlasi* in Serbia and Bulgaria. To the Vlach themselves, they are known as *Aroumāni*, *Armâñi* or *Rrămăñi* (Kahl, 2002, p. 145; Winnifrith, 2002, p. 121).

The Vlach ethnonym appears for the first time in 11th century AD Byzantine sources describing events in Greece, though an even earlier date can be inferred in association with two episodes that took place in the territories around Kastoria and Thessaly during the years 976 and 979 (Wolff, 1949, p. 204; Kaldellis, 2019, p. 240). From the second half of the 12th century AD, the name Vlach also emerges as a geographical label, such as Wallachia or 'Great Vlachia' in the Thessalian highlands, 'Little Vlachia' in Aetolia-Acarnania and 'Upper Vlachia' in Epirus (Soulis, 1963, pp. 271–273; Osswald, 2007, p. 129). Their ethnic origin has long puzzled scholars and several hypotheses have been drawn, with some sharing the assumption that Vlachs are the descendants of Romanized local Balkan tribes who settled in isolated highland pockets during the Slav penetration in the Early Middle Ages (Wace & Thompson, 1914, pp. 265–273; Winnifrith, 1987, pp. 88–89). Historical sources often depict Vlachs as transhumant shepherds (Gyóni, 1951, pp. 29–41), although there are also

DOI: 10.4324/9781003422952-14

instances when they appear as muleteers, keepers of passes, guides and tradesmen, with the distinguished pattern of being 'always on the move', as an aid to their remarkable survival (Winnifrith, 1987, pp. 71–72, 105–106, 110–112, 125, 148).

The Upper Vjosa Valley marks the northern border of the historical region of Epirus and played an important role in connecting the hinterlands of central Albania and the Ionian and Adriatic coastlines with northern and central regions of Greece and western North Macedonia (Soustal, 1981, pp. 88–89). Given this position, the valley was one of the main corridors of Vlach nomadic movement, from the winter pasture lands of coastal Ionian plains to the summer pastures of the immediate valley uplands and further east in the districts of Kolonja and Korça. These territories record a considerable presence of Vlach settlers, likewise the neighbouring mountainous area of Pindus in Greece to the south of the valley.

There are several stages of Vlach settlement in the Upper Vjosa Valley, stemming from the first record in the early 15th century AD, until their final settling at the end of the 1950s. The Vlach historical trajectory was not an easy one, but ultimately, they managed to preserve their language, culture and identity. It is unfortunate that in more recent years, these distinctive ethno-cultural features have been threatened with an imminent risk of extinction. It was this potential loss that drew our attention and directed the focus of the research in the Vjosa Valley towards capturing the last surviving memories of Vlachs who once conducted a transhumant way of life. They provided significant information about their temporary settlement locations and organization as well as their long-distance nomadic routes. The mobile way of life required Vlachs to temporarily reside in an area during the summer and winter pastoral seasons, where they established a domestic environment consisting of single family hut dwellings known as *kalive*. While these would perhaps appear to some as 'simple huts', this would contradict the complex conception of space that characterizes the Vlach way of life. As such, the *kalive* is a traditional building, no less important than a stone-built structure. Archaeological studies have proven that communities have lived in ephemeral dwellings since prehistory. Today, these structures are widely reconstructed experimentally as a means of testing an archaeological interpretation, as well as creating heritage landmarks for education and presentation purposes.[1] In the Vjosa Valley, we were limited by the absence of archaeological data and therefore, we used surviving memories to build a Vlach hut dwelling site, and eventually recreate a lost symbolic aspect of their heritage. Traditional data can be used to interpret the past similarly to scientific archaeological experiments (Stone & Planel, 1999, pp. 1–7). The difference, however, stands in the type of validated data collected. In our case, the goal was to validate information gathered from an ethnographic survey through a practical construction experiment, while also trying to recreate and interpret an aspect of Vlach past. In so doing, it also rescued a traditional knowledge that is fast disappearing. Past memories were used equally to record, design and map the nomadic routes, which were assessed and surveyed on the ground, so as to preserve the historic practice of transhumance and create a visual heritage product.

## 14.2   Vlachs in the Upper Vjosa Valley

Today, Vlachs are becoming increasingly less evident in the southern Balkans, and there are obvious reasons for this disappearance. To begin with, the loss of transhumance, which had been the source of the Vlach's survival for many centuries, encouraged a movement towards large urbanized areas, where it was difficult to preserve a minority identity (Hammond, 1976, p. 46; Winnifrith, 2002, pp. 112–121). Additionally, the establishment of one-language national schools and the abandonment of endogamy further accelerated the integration and assimilation process of the Vlachs. This is well reflected in the Upper Vjosa Valley, where many of the Vlach inhabited villages located in isolated upland areas are now abandoned and there are only a few among the descendants of the Vlachs who are aware of their identity and even fewer who can speak the language.

The assimilation process of Vlachs in the Balkans seems to have begun with the establishment of independent nation-states during the 19th and early 20th centuries. The formation of new states led to their geographical areas of origin being divided by new national borders. Although it was impossible to develop a state of their own, many among the educated Vlachs in Albania became prominent figures of the country's national movement, as did their counterparts in Greece. At the same time, the Romanian state also began influencing activities that encouraged an independent Vlach identity. They used the proximity of the Vlach language to Romanian, sending missionary teachers to establish Aromanian schools in Macedonia, Thessaly and Epirus (Kahl, 2003, pp. 208–214). Such schools were also set up in the Vjosa Valley, in the town of Përmet and the village of Kosina, which may have influenced the preservation of the Vlach identity amongst their pupils.

In the aftermath of the Second World War, the communist regime in Albania encouraged the rise of a fiercely nationalistic identity and the formation of a 'New Man' for their new, modern society. In this context, the traditional Vlach way of life and their socio-economic organization were perceived as relics of the past that belonged to an era of 'medieval slavery' (Prifti, 1971, pp. 12–13). Furthermore, the sealing of national borders between Greece and Albania prevented pastoral mobility across the two countries and the Albanian reforms for state-led collectivization banned Vlach pastoral nomadism, forcing them to settle permanently. However, Albania's isolation and control of its internal population movement somehow aided the preservation of the Vlach identity. Following the collapse of the communist regime in 1991, the country endured a long political, economic and social crisis which created an unfavourable climate for the survival of a minority identity. During these years some Vlachs attempted to seek a new Greek or Romanian nationality, whereas others retained their *Vlachness* and still remained loyal to the Albanian home state.

Another potential threat to the survival of the Vlach identity is a disturbing persistence of prejudice. This is likely associated with a general perception of them as shepherds, following a 'primitive' nomadic way of life with temporary hut settlements. It was this particular lifestyle that led the sedentary Vlachs in the Vjosa Valley to differentiate themselves from their nomadic fellows. Moreover, it is not unlikely that these feelings of intimidation may have been cultivated continuously; indeed, since the earlier appearance of the Vlachs in historical sources, they are often described as 'faithless, infidel, perverse, untrusted, barbarians, brigands, etc.' (Kaldellis, 2019, pp. 241–242).

The Vlachs in the Vjosa Valley (district of Përmet) distinguish themselves into two groups. The first is a large group of earlier dwellers known by Albanians as *Çoban*, a term loaned from Turkish for shepherd, but they generally call themselves *Aromani*. The second, a smaller group, locally known as *Vllah* and among themselves as *Rrmani/Ramani* have conducted a semi or fully nomadic way of life until the end of the 1950s when they were settled permanently in the territory.

Historically, the earlier Vlach residents inhabited the highland regions of Dangëlli and Shqeri, expanding from the lower valley towards the east and centred mainly around the village of Frashër. During the 17th and 18th centuries, this village was a large rural settlement and a trading post along the route between the Vjosa Valley, the district of Kolonja and the medieval town of Voskopoja (*Moschopolis*) in the Korça district, which was once the most famous Vlach settlement in the Balkans (Adhami, 1998). A smaller presence is recorded in the villages along the western flank of the Vjosa Valley, at the foot of mountains Dhëmbel and Nemërçka, including the town of Përmet (Figure 14.1).[2]

Those Vlachs living in villages were mainly shepherds of their flocks or hired herdsmen, muleteers and excellent charcoal makers, whereas those in city were craftsmen such as tailors, blacksmiths and tradesmen.

There is an assumption that it is from the village of Frashër that the name of one of the largest Vlachs groups of the southern Balkans originates. These are known as *Farsherot*, or by the Greeks as *Arvanitovlakhi*, and are to be found in central and south Albania, Greece and North Macedonia (Wace & Thompson, 1914, pp. 206–218, 223). Some of the later Vlach residents in the valley also claim a *Farsherot* origin, whereas others say they have come from different areas in northern Greece, particularly from the village of Kefalovrison, located at the southern end of Mount Nemërçka, close to the Albanian border (Figure 14.3). The Vjosa Valley was a regular seasonal nomadic route for many Vlachs, and once it became their permanent home, they settled in the villages of Kosina, Kutal, Badëlonja and Lëshica. The first three of these villages have continuous historical records of Vlach presence, whereas the latter has no earlier evidence, but was a camping ground along their transhumance route.

There is no point in defining ethnic and cultural boundaries between the earlier *Çoban* residents and the later settled Vlachs. This is a present-day

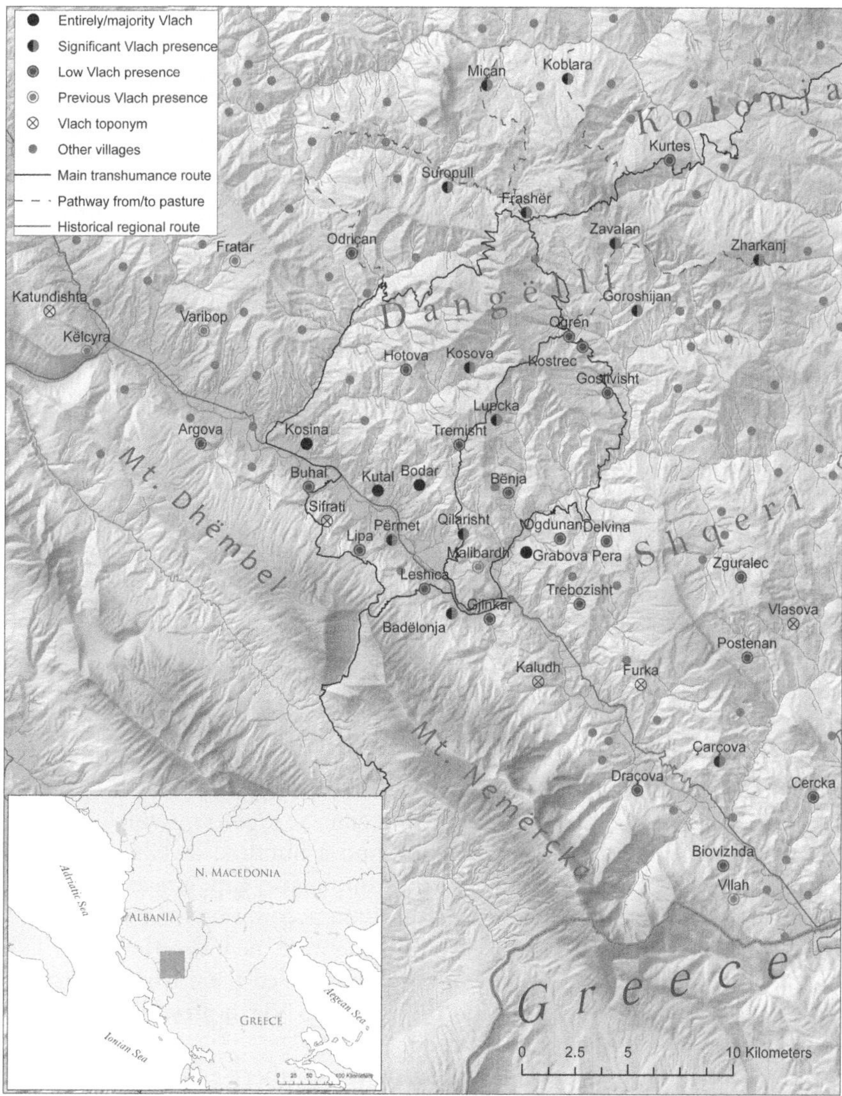


*Figure 14.1* Map of settlements with Vlach presence in the Upper Vjosa Valley.

social distinction, particularly amplified when the latter group becomes permanent residents of the valley, at the end of the 1950s. Both groups speak the same dialect, are conscious of their Vlach identity, confess to Eastern Orthodox Christianity and consider the Upper Vjosa Valley a homeland or a main hub of their nomadic movement. It is likely that over time, transhumant Vlach colonies settled in the valley and either established new settlements or formed hamlets on the edges of existing villages, most probably where other Vlachs resided. The pattern of constant movement is reflected by a mid-19th-century census of the Eparchy of Përmet, which records village families based on their Christian and Muslim religions (Aravantinos, 1856, pp. 370–375). Some of the villages traditionally inhabited by Vlachs, such as Bodar, Kostrec, Zavalan, Miçan and Suropull are listed in this census as only having Muslim families, inferring there were no Christians and therefore no Vlach habitants. Earlier in that century, however, Vlachs were reported in Kostrec and Christian families in Bodar (Pouqueville, 1826, pp. 265, 270). A particular type of movement may be deduced for settlements along the western side of the valley, where Vlachs are generally smaller in number. Here, newcomers are in a continuous flow, but only for a brief temporary stay. For example, the village of Buhal always had few Vlach families who only stayed for a generation and worked as hired herdsmen, practicing year-round pastoralism with short-distance movements on Mount Dhëmbel. Later, when a family left to settle elsewhere, i.e., in the village of Draçova to the south, others came from the village of Fratar to the north. The settlements in the uplands to the east, unlike those in the west, have greater population numbers and longer staying migrants. This may be due to extensive rich summer pastures, a more independent economy, as well as being located along main transhumance routes with other Vlach centres of Kolonja district and Voskopoja nearby to facilitate a constant population flow and presence.

## 14.3    Tracing the past

Earlier evidence for a Vlach presence in the Vjosa Valley comes from the 1431 AD Ottoman cadastre of Këlcyra (medieval *Kleisura*), which mentions the town as having a Vlach population and a hundred dwellings (Soustal, 1981, p. 182). Subsequently, there is a chronological gap of nearly 400 years until the Vlachs briefly re-emerge in a written account by the French traveller François Pouqueville (1826, p. 270), where he mentions that the village of Kostrec was inhabited by Vlachs who had moved there from Voskopoja. This does not mean, however, that there was no Vlach presence during this interlude or prior to their first mention.

Këlcyra occupies a dominant position at the northern extreme of the Upper Vjosa Valley, overlooking roads from the west and northeast. The archaeological and historical data suggests that the site was an important fortified centre with occupation phases dating from Hellenistic to Medieval times. Outside the fortification, there are stone-built houses recorded over an area of *c*.9.80ha which were likely part of a nucleated settlement, developed from the 10th to 11th centuries AD with a later expansion during the 13th to 14th centuries (Miti et al., forthcoming); these houses may be associated with those mentioned in the Ottoman cadastre. In late 13th-century historical accounts, the town appears as an administrative and commercial Byzantine centre, which soon afterwards was under the control of Albanian lords with Venetian merchants also residing there (Xhufi, 2009, pp. 176–177). It is, therefore, doubtful to assume that Këlcyra was entirely inhabited by Vlachs for in the late Byzantine sources that described events from the region of Epirus, Vlachs appeared to be coalesced or confused with Albanians (Osswald, 2007, p. 130). The early 19th-century traveller, Thomas Hughes (1820, p. 272) recorded an oral history describing the settlement being abandoned half a century earlier, and that the people had migrated to the neighbouring mountain region of Kolonja. It is not certain whether these were Vlachs, but their recorded shift towards Kolonja, an area of Vlach habitation, and their tradition for moving to places where other Vlach resided, may infer they had a presence in Këlcyra.

The Ottoman cadastre of 1435 AD for the *vilayet* (province) of Përmet (Caka, 2018) contains no direct accounts of Vlachs, although some of the place names listed may indicate a Vlach presence at that time or even earlier. For instance, the village of *Vllaşova/Ulasova* that could be identified as *Vllás-ovon* (present Lashova) mentioned in the mid-19th century by Aravantinos (1856, p. 372) is clearly an ethnonym deriving from the Slavic-Bulgarian name *Vlas* denoting Aromanian (Ylli, 2000, p. 188). This may suggest an earlier use of the name, perhaps from the time, the territory was under the Bulgarian domination during mid-9th and 10th centuries AD. The same origin and explanation may be implied for the village of *Vllah*, located at the border with Greece, which appears in the early 19th century as a hamlet inhabited by shepherds (Pouqueville, 1826, p. 253), and a century later as having three Vlach families living there (Burileanu, 1912, p. 358). Today, the village is inhabited by Greek-speaking minorities and it remains unclear whether the then-settled Vlachs moved elsewhere or were absorbed by the local majority.

There are a number of place names with Latin origin that emerge from the Ottoman registers, including the town of Përmet (*Premeti*), and it is tempting to suggest that the Vlach language may have been responsible for their survival. Among these Latin toponyms, however, only a few may be directly related to the Aromanian language, such as *Frata*, *Fratan* (Fratar) and *Sifrati* (Sfrat), which are probably from the word *frater* meaning 'brother/brotherhood', and also *Furka* (Furkëza) from *furca* meaning 'distaff'. Of all these place names, only Fratar, now hosting an exclusively Albanian population had previously evidence of Vlach presence. *Sifrati* and *Furka* were previously abandoned and there is no information about their ethnic composition, while *Frata* remains as yet unidentified. Other village names, such as *Kaliuvi* (Kaludhi) and *Katunişte* (Katundishta, located at the outskirt of Këlcyra), may perhaps have once been associated with Vlach settlement. These place names may derive from the Aromanian words *kalive/călívă* meaning 'hut' and *katun/cătùn* meaning 'hamlet', although both originated or were borrowed from other southern Balkan languages.[3] We should also consider that in the Albanian lexicon there are many words of Latin origin. Indeed, it would be a mistake to use the toponomy as a sole argument for interpreting Vlach history in the Vjosa Valley, for as many present-day Vlach-inhabited villages have a non-Latin toponymy, namely of an Albanian or Slavic-Bulgarian origin.

Archaeological research in the Upper Vjosa Valley suggests a fairly densely occupied Roman and early Byzantine (mid-2nd century BC to the 6th century AD) rural landscape, with sites located in the lowlands along the main route and Përmet acting as a central place (Miti & Serjani, 2023). It is perhaps from this time that one has to look for the origin of many of these Latin place names. During the 9th and 10th centuries AD, when the territories were under Bulgarian dominance, archaeology has detected traces of occupation in the fortified strategic sites of Këlcyra, Përmet and Petran, suggesting an increased interest over the main route through the Vjosa Valley, and the revival of a secondary artery that started from Petran and ran east towards upland territories. This is also a section of the main transhumant Vlach route. It is thus not impossible to surmise that many of the upland settlements with Slavic names, such as *Vllaşova/Ulasova* may have been founded at this time, or if previously inhabited were only to a small degree. Some scholars have assumed that there were Vlachs hidden behind or mingled with the Bulgars, and the latter may have perhaps somehow aided the survival of the Romance language (Wace & Thompson, 1914, pp. 272–273; Winnifrith, 1987, p. 100). Furthermore, the location of known Roman sites do not correspond with the present-day upland Vlach-inhabited settlements, with the exception of Kosina, Kutal and Bodar, whose inhabitants claim to have migrated from the region of Dangëlli. On the other hand, it is impossible to assume the presence of 'Romanized' pastoral communities in these upland territories for as long as archaeological evidence is absent. Besides, there is a long chronological gap to bridge from the time the Vlachs were first mentioned in the Vjosa Valley and any surviving group of Latin speakers in the Early Middle Ages.

Any attempt to associate archaeological artefacts or architectural building remains found in the Vlach-inhabited villages with a particular ethnic group could lead to a hasty and erroneous interpretation. For example, the 11th to 12th century AD silver bracelets found as grave goods in the village of Ogren are of a typical Byzantine fashion (Anamali, 1980, p. 11, n. 24); while the early 13th century AD church at Kosina represents a regional architectonic style widespread in the Despotate of Epirus (Koch, 2006, p. 43). The same may be said for the mid-16th and late 17th century churches of Tremisht and Ogren, whose ground plan and wall painting style are similar to other churches in the region (Thomo, 1998, pp. 89, 236). Equally, the Vlach stone-built houses in the village of Lupcka are no different to the houses of Albanian co-inhabitants, or indeed from traditional dwellings in other villages of the valley.

## 14.4   Reconstructing a lost heritage

As yet, it is unclear what the ancient Vlach settlement looked like with history and archaeology unable to fully assist in this descriptive regard.

However, if we assume that Vlachs conducted a nomadic way of life for many centuries, then it is fair to say that their traditional settlement resembled the hut-encampments recorded in early 20th-century accounts. Traditional dwellings are usually considered as a built-form of expression, usually of common people, that once adopted as the norm of a particular society is then transmitted through generations (Bourdier & Alsayyad, 1989, pp. 5–6). In so speaking, the Vlach hut-dwelling, the *kalive*, may be seen as a traditional type of architecture, whose form and symbolism were the result of a transhumant way of life that called for an entire family to become part of a continuous seasonal journey. It represents a particular type of built heritage that demonstrated the adaptability of Vlach builders, who made do with whatever the local environmental had to offer, yet also managed to display notable cultural indicators. The *kalive* dwelling was made of perishable materials, meaning no ground remains are left for documentation purposes. Also, Vlach builders did not design plans for their houses, and yet the construction knowledge persisted to be passed down between generations. Some of the Upper Vjosa Valley Vlachs were once part of mobile pastoral societies who relied on *kalive*-building knowledge to house themselves during seasonal movements across the landscape of southern Albania up until the mid-20th century. During our field survey, many elderly Vlachs were interviewed and provided essential information about various aspects of their past transhumant life, including the nomadic movements, settlement settings and organization, as well as many details of the *kalive* construction. However, when interviewing people about the past, in the context of a contemporary setting, there may be a lack of consistency between what is told and what might really have happened, which can result in a fragmentary type of data that limits the understanding of the past (Hodder, 2012, p. 43). This necessitated us to gain some verification of the collated word-of-mouth evidence in order to present a reasoned interpretation of the Vlach past. It was at this point that the project undertook an experimental reconstruction to verify these last surviving first-hand memories of the Vlach people regarding their historical dwelling. The building process was particular as it followed the lead of the traditional builders who retained a detailed memory of the *kalive* construction and who, once in contact with the chosen material, had the ability to work these with methods widely used in the past. By involving previously active *kalive* builders, it enabled us to document the construction techniques and materials of this specific building in the hope that this information may also be used as a reservoir of ideas to inform wider studies. A constructed heritage object is particularly important because of its extraordinary power to attract the wider public (Blockley, 1999, p. 16), and therefore it offers more opportunities for the unknown historical narrative to be communicated. Furthermore, it offered the local Vlachs an opportunity to appreciate their culture, rediscover their past and develop a sense of belonging to the Vjosa Valley territory.

### 14.4.1   Construction of the kalive

The *kalive/căliva* was the focal point of the Vlach encampment.[4] It housed a nucleated family with many domestic activities being undertaken there, mainly wool processing which was a female-dominated endeavour. An encampment often consisted of 20–30 hut dwellings, although in earlier times, fifty to one hundred of these were estimated to have existed (Gyóni, 1951, p. 31). Sometimes, summer encampments retained a fixed location for two or more years and if dwellings withstood the temporary abandonment, they could be repaired for use with little effort. In many cases, the *kalive* was also the winter dwelling, meaning it was the primary year-round residence. Historical evidence regarding Vlach dwelling is scarce and limited only to a general depiction of their rounded form and the interior layout (Burileanu, 1912, pp. 114–116; Hammond, 1967, p. 25). A better descriptive account is given by the Albanian scholar Koço Zheku (1973, pp. 81–83), who not only details the construction techniques used, but also attempts to regard it as a traditional rural dwelling.

The new *kalive* construction site was set on the outskirts of Qilarisht village which was once a temporary station along the Vlach's transhumance route. The entire construction process, including the *in situ* material collection, involved the work of seven people for 13 days. As many years had passed since the leading elderly Vlach master had constructed his last *kalive*, this process took a longer time than it would have done in the past. Thus, certain construction stages had to be repeated and more time was dedicated to learning and transferring key skills or building techniques. The process involved basic working tools, including axes, crowbars, saws, hammers, sickles, scissors, string, wire, etc.[5]

It was initiated by drawing a circle of 5m in diameter with a wooden stick tied to a string measuring the radial length set on a fixed wooden pole at the centre point of the hut. Next, 30 Ash (*Fraxinus* sp.) poles of c.0.05–0.08m in diameter and 2.50–3.50m long were set into post-holes dug at 0.40m intervals around the perimeter, leaving a space of 0.80m for the doorway. For the roof, 30 poles of 0.04m in diameter and over 3m long were tied around a 0.45m diameter wooden-ring made of a thin cornel branch. This was placed at the top of a wooden beam at a height of 4.60m and moved at the centre of the hut, remaining there for structural stability until the end of the construction process. The ends of the roof poles were then joined with the upper ends of the side wall poles. These vertical poles were secured on the outside with thinner horizontal branches (80 pieces) set every 0.25m from the base upwards to form the skeletal frame for the building (Figure 14.2a). The process of covering the frame began by placing bunches of ferns along the outside walls, secured using horizontal withies lashed on with wire (Figure 14.2b). The roof was covered with straw and bulrushes using the same techniques as for the walls (Figure 14.2c). The inner walls were covered with tightly set reeds of 0.80m in height from the ground level to make a solid base upon which a mud-plaster render could be keyed onto. Inside, a hearth was placed

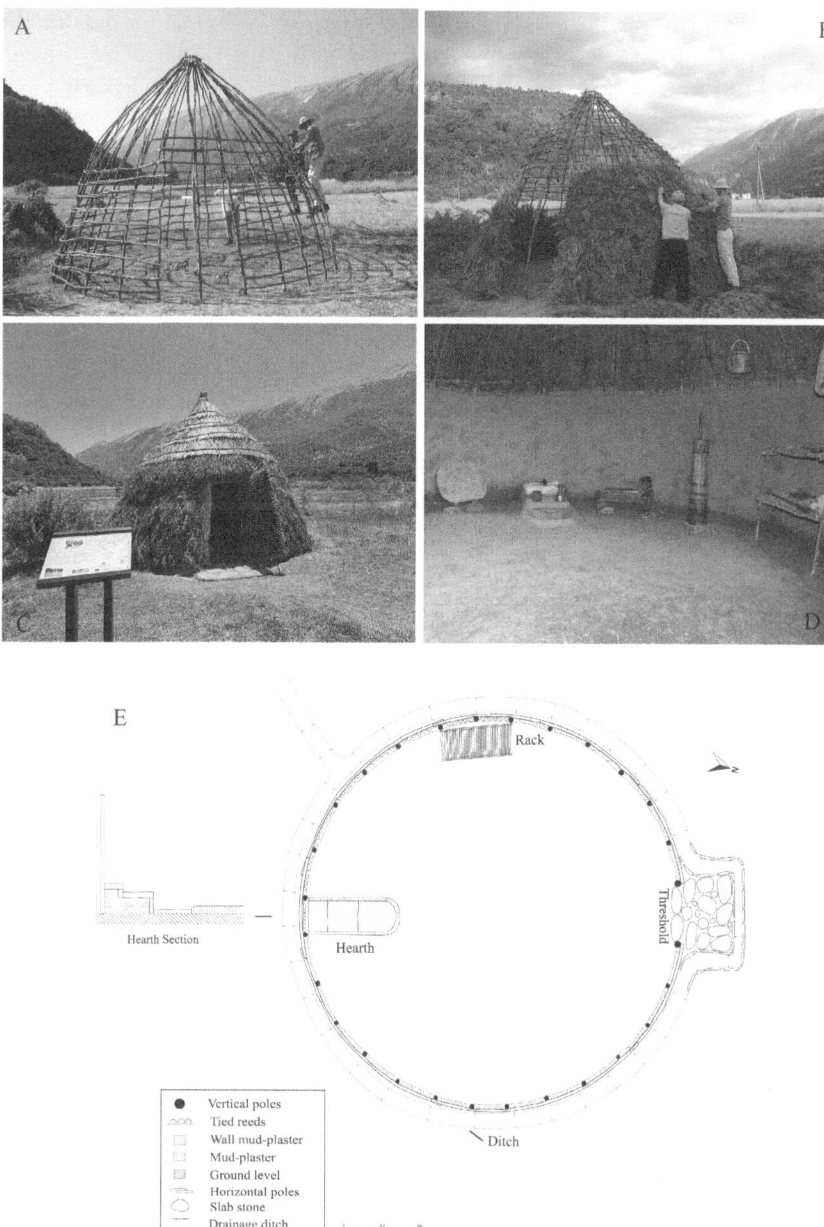



*Figure 14.2* Construction of the *kalive* dwelling site.

opposite the doorway which consisted of a firepit lined with tufa stones and two stone-covered upper platforms. In addition, a two-shelf rack made from string-lashed rods was set along one side of the inner wall. The next step was to complete the floor and render the interior walls. The plaster for this was prepared by mixing clay, water, straw and manures, which was heavily treaded to form a compact and gooey mass. This was then hand-thrown and smoothed on the walls and floor of the dwelling. Once dried, several layers of diluted plaster were applied to cover any cracks and give it a neat finish. The door was made of woven cornel branches and attached to one side of the doorway pole. Finally, a drainage ditch was dug around the outside of the hut and a stone slab threshold was set in front of the doorway (Figure 14.2d–e).

Additionally, the building's interior was furnished with a selection of essential, everyday items from a previous nomadic life, giving it a more captivating and engaging museum-like quality. The results of a questionnaire administered onsite, as well as ongoing communication between staff members and visitors, demonstrate that the recreated dwelling and the inherent complexity of Vlach nomadic life as a whole represent a cultural heritage asset of particular interest to the local cultural tourism sector.

### 14.4.2   *The transhumance routes*

For as long as the Vlach practised long-distance transhumance, each late April pastoral family groups were taken on a migratory journey accompanying a caravan of livestock movement from lowland pastures towards grazing lands in the high mountains, returning the same way at the end of October. The principal itinerary of this seasonal nomadic movement started from the Ionian coast towards the southwest on both sides of the Greek-Albanian border and continued northeast across the Drino Valley and further east through the Dhëmbel mountain pass. From here, just above the town of Përmet, the route diverted into three alternative branches that ran eastwards across the fertile pastures of the Dangëlli region to rejoin at Frashër. The route then continued further east to Mount Gramoz as well as northward to the pasture lands around Vithkuq and Voskopoja.

The journey lasted from two to three weeks over a distance of *c.*220km, depending on the weather, local opportunities to trade goods or the emerging rich grazing that may delay flock movement. Along the way, they resided in daily camps of woolen tents. Upon arrival at the summer pasture site, the Vlachs built their hut-encampments which were normally located at the periphery of inhabited areas and close to water sources. It is not surprising therefore that once the opportunity arose, these temporary camps grew to become permanent quarters of the nearby settlements. Other long-distance pastoral routes may also have passed along the Vjosa Valley, such as those coming from the Myzeqe plain summer pastures on the Adriatic coastline to the northwest, or from the lowlands and mountainous pastures to the south in Greece.

These itineraries were compared with the available historical information and processed with the aid of GIS. This enabled to predict and map secondary or alternative pathways that lead to pastures near the

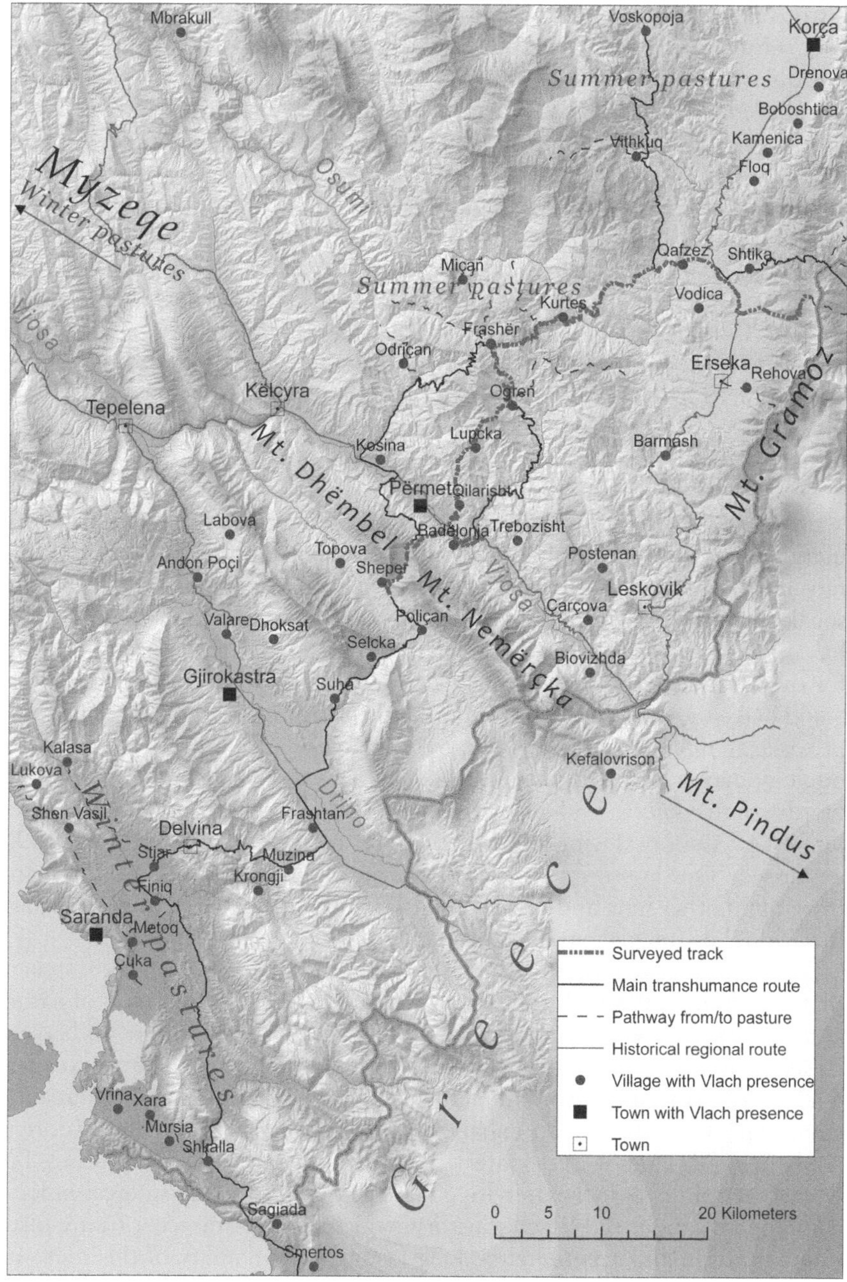


*Figure 14.3* Map of the transhumant routes and the main centres with Vlach presence in the region.

historical Vlach-inhabited settlements. Furthermore, a section of the route was subsequently surveyed to assess the preservation and viability of the track as well as evaluate its future potential use as a heritage trail. The surveyed section begins at the Dhëmbel mountain pass and runs towards the upland region of Dangëlli, ending up on the western slope of Mount Gramoz (Figure 14.3). The hike lasted for five days covering a distance of 97km and enabled us to collect information about traditional villages, monuments and other historical landscape resources along the way. The route was recorded using GNSS and presented as a digital online map,[6] which can be accessed by scanning QR codes included in an information panel at the *kalive* dwelling site and in the recently launched tourist leaflets. The online accessibility of the route is important as it allows visitor numbers to be monitored, and an evaluation of its use as an emerging tourist destination.

## 14.5    Concluding remarks

When considering the creation of a heritage asset that embodies an ethnic marginalized group, the expectation of receiving support from that community itself is quite low. It was not an easy task to convince the Vlachs to speak about their past life, which is understandable since it is an aspect that has differentiated them from other parts of society. The research approach for overcoming this initial hurdle was to generate a historical narrative that could make the community feel proud of their history, with the hope it would provoke an interest and reveal past memories. For such initiatives, this methodology may be the route to success. This was clearly demonstrated by the words of the *kalive* lead master, who while experiencing the unexpected interest of visitors during the experimental reconstruction, enthusiastically expressed: 'Our house was simple but at the end that was our tradition, and if we build other huts and preserve our language, we have created a heritage that would distinguish and represent us today'.

The reconstructed dwelling site is located in a previous Vlach encampment that at present is a camping site area associated with the promotion of nature and cultural tourism, which has therefore enabled the Vlach hut to become an attractive focal point for visitors. Also, the nomadic trail route concept, while preserving and promoting the historical practice of transhumance, could equally contribute to enhancing the visibility of the living historical landscape. We cannot anticipate what will happen in the near future, but what is certain is that this initiative managed to activate an interest in local communities about the Vlach's history, which may guarantee a future institutionalization and eventuate the long-lasting sustainability of this minority heritage.

The peaceful survival of the Vlachs in an ethnically troubled Balkan society is remarkable, and this deserves recognition. During the last century, the Vlachs were located in disputed borderland territories, but they were never a potential source of any major ethnic conflict as has been the case with other ethnicities in the Balkans. On the contrary, they co-existed peacefully and contributed to the progress of the nations in which they lived, making them a good example of how an ethnic minority can aid in bridging the gap between national divergences and social-cultural diversities.

## Acknowledgements

We thank all those who supported the project in the Vjosa Valley, in particular the Vlach *kalive* lead master Mihal Kuro, his son Jorgo, and project team members Ervis Kasa, Renald Meta and Miklevan Deliu. We are also grateful to Emily Glass and David Bescoby for proofreading the paper.

## Notes

1 For reconstructed prehistoric huts in England, see Townend, 2007, pp. 97–111; see also other examples in England, Denmark, Germany, and etc., in Stone and Planel (eds.), 1999.
2 Our survey data combined with late 19th and early 20th century historical sources recorded a Vlach presence in 43 villages and hamlets (Burileanu, 1912, p. 392; Adhami, 2001, pp. 53–54; see also a map by Thede Kahl published in www. farsharotu.org/vlach-map-of-albania).
3 The word *kalive* probably originates from ancient Greek *kalúve, kalívion* meaning hut/tent, appropriated in Albanian and southern Slavic as *kolibe*. The word *katun* appears for the first time in the Byzantine sources as *katouna* to denote a Vlach mobile hamlet, and was later adopted in Albanian as *katund* and in the southern Slavic as *katun* to describe a previous hamlet that has become a permanent village (Ajeti, 1976, p. 108; Kaldellis, 2019, p. 241).
4 The term *kalive* is generally associated with dwellings in temporary encampments, although there are cases, as, for example, the Vlachs of Mount Pindus, who use the word *kasă* for both a hut and a stone-built dwelling (Wace & Thompson, 1914, p. 99), or the *Farsherots* of southern Albania who distinguish between *kalive*, a dwelling in summer camps, and *căsă* a winter house, which probably suggests a semi-nomadic way of life.
5 The tools are the same as those used in the past, except for the steel wire utilized for lashing; they used brush broom (*Spartium junceum*) branches in traditional construction, but due to the limited time and the need for assuring a long-lasting stability of the structure, wire was used instead.
6 https://www.wikiloc.com/hiking-trails/vlachs-transhumance-route-1086 59966.

## Bibliography

Adhami, S. (1998). *Voskopoja dhe monumentet e saj*. Tirana: Infbotues.
Adhami, S. (2001). *Vështrim mbi Kulturën Popullore të Trevës së Përmetit*. Tirana: ADA.
Ajeti, I. (1976). Pour servir à l'histoire des anciens rapports linguistiques albano-slaves. *Iliria* 5, 105–112.
Anamali, S. (1980). Antikiteti i vonë dhe mesjeta e hershme në kërkimet shqiptare. *Iliria* 9–10, 5–21.
Aravantinos, P. (1856). *Chronographia tis Ipeirou, Vol. II*. Athens: Typ. S.K. Vlastou.

Blockley, M. (1999). Archaeological reconstruction and the community in the UK. In P. Stone and P. Planel (eds.), *The Constructed Past: Experimental Archaeology, Education and the Public*, 15–34. London, New York: Routledge.

Bourdier, J. P. and Alsayyad, N. (1989). Dwellings, settlements and tradition: A prologue. In J. P. Bourdier and N. Alsayyad (eds.), *Dwellings, Settlements and Tradition. Cross-Cultural Perspectives*, 5–25. Lanham, MD: University Press of America.

Burileanu, C. (1912). *I Romeni di Albania*. Bologna: Reale tip. cav. L. Andreoli.

Caka, E. (2018). *Defteri i Regjistrimit të Përmetit dhe Korçës i vitit 1431*. Tirana: ASA.

Dvoichenko-Markov, D. (1984). The Vlachs: The Latin speaking population of Eastern Europe. *Byzantion* 54, 508–526.

Gyóni, M. (1951). La transhumance des Vlaques Balkaniques au Moyen Age. *Byzantinoslavica* 12, 29–42.

Hammond, N. (1967). *Epirus*. Oxford: Clarendon Press.

Hammond, N. (1976). *Migrations and Invasions in Greece and Adjacent Areas*. New Jersey: Noyes Press.

Hodder, I. (2012). *The Present Past. An Introduction to Anthropology for Archaeologists*. South Yorkshire: Pen & Sword Books.

Hughes, T. (1820). *Travels in Sicily, Greece and Albania. Vol II*. London: Printed for J. Mawman.

Kahl, Th. (2002). The ethnicity of Aromanians after 1990: The identity of a minority that behaves like a majority. *Ethnologia Balkanica* 6, 145–169.

Kahl, Th. (2003). Aromanians in Greece: Minority or Vlach-speaking Greeks? *JKGS* 5, 205–219.

Kaldellis, A. (2019). *Romanland: Ethnicity and Empire in Byzantium*. Cambridge, MA: Harvard University Press.

Koch, G. (2006). Einige Bemerkungen zur Kirche in Marmiro/Albanien. *Deltion tes Christianikes Archaiologikes Hetaireias* 27, 43–48.

Miti, A., Serjani, E. and Civantos, J. M. (forthcoming). Medieval ceramics from the Upper Vjosa Valley, Southern Albania. In *Proceedings of XIIIth AIECM3, Granada, 8–13 November 2021*.

Miti, A. and Serjani, E. (2023). Transformations of a post Roman countryside: First results from field survey and excavations in the Upper Vjosa valley (SE Albania). In I. Borzić et al. (eds.), *TRADE - Transformations of Adriatic Europe (2nd – 9th Century). Proceedings of the Conference, Zadar 11th–13th, February 2016*, 93–103. Oxford: Archaeopress.

Osswald, B. (2007). The ethnic composition of medieval Epirus. In S. Ellis and L. Klusáková (eds.), *Imagining Frontiers, Contesting Identities*, 125–154. Pisa: Edizioni Plus.

Prifti, N. (1971). Nomadët e Fundit. *Revista Ylli* 12, 12–13.

Pouqueville, F. (1826). *Voyage de la Grèce*. Paris: Firmin Didot 1826–27.

Soulis, G. C. (1963). The Thessalian Vlachia. *Recueil des Travaux de l'Institut d' Études Byzantines* 8(1), 271–273.

Soustal, P. (1981). *Nikopolis und Kephallenia, Tabula Imperii Byzantini 3*. Wien: Österreichische Akademie der Wissenschaften.

Stone, P. and Planel, P. (1999). Introduction. In P. Stone and P. Planel (eds.), *The Constructed Past: Experimental Archaeology, Education and the Public*, 1–14. London, New York: Routledge.

Thomo, P. (1998). *Kishat Pasbizantine në Shqipërinë e Jugut.* Tirana: Kisha Orthodokse Autoqefale e Shqipërisë.

Townend, S. (2007). What have reconstructed roundhouses ever done for us..? *Proceedings of the Prehistoric Society* 73, 97–111.

Wace, A. and Thompson, M. (1914). *The Nomads of the Balkans.* London: Methuen & Co. Ltd.

Winnifrith, T. (1987). *The Vlachs: The History of a Balkan People.* London: Duckworth.

Winnifrith, T. (2002). Vlachs. In R. Clogg (ed.), *Minorities in Greece*, 112–121. London: C. Hurst & Co. Ltd.

Wolff, R. (1949). The 'Second Bulgarian Empire': Its origin and history to 1204. *Speculum* 24(2), 167–206.

Xhufi, P. (2009). *Nga Paleologët te Muzakajt. Berati dhe Vlora në shekujt XII–XV.* Tiranë: Shtëpia botuese '55'.

Ylli, Xh. (2000). *Das slavische Lehngut im Albanischen.* München: Verlag Otto Sagner.

Zheku, K. (1973). Të dhëna mbi kasollet fshatare të vendit tonë dhe gjurmët e tyre në lashtësi. *Monumentet* 5–6, 81–100.

# 15 Community genealogy as a tool for heritage tourism

*John Tierney, Maurizio Toscano and Amanda Slattery*

Since the development of the World Wide Web, genealogy has evolved into a key element of multidisciplinary historical investigations of human mobility and migration, name studies (onomastics), micro- and macro-histories as well as amateur family histories. This chapter concentrates on an approach to genealogy developed by Irish communities and a team of Irish archaeologists using the web platform historicgraves.com, where good quality gravestone survey datasets, recorded in the field, are published and enriched, as a collective citizen-science project. The individual benefits of the increase in digital availability of historical datasets are obvious but there are wider community and national benefits to these ostensibly individual surveys. National and international benefits are recognised and supported at different levels within different countries, particularly in those territories with strong associated diaspora populations.

Irish community approaches to genealogy offer interesting lessons in communicating with migrants, understanding the development of diaspora populations across generations and also assisting in identifying common elements with other ethnicities. We differentiate local community-led heritage projects as a subset of citizen science projects, on the basis that a group of unrelated, place-based individuals undertake a particular project for a broader benefit. Community-led genealogy in Ireland received a short-term boost with the 2013 Gathering Ireland project (Miley, 2013). The Gathering was a national campaign which had the foresight to give preference to a bottom-up approach with local communities, encouraged and supported to deliver a wide range of genealogical projects, which constituted 56% of all events organised (Tottola, 2018). Our own Historic Graves Project was recommended as part of the subsequent Diaspora Toolkit (Kennedy & Lyes, 2015) along with the Ireland Reaching Out project, which also has a strong community volunteer element built on top of a web platform.[1]

Communication with heterogeneous diaspora populations is challenging in both the home and destination countries but it is achievable. The Irish diaspora is one of the most developed of global diasporas and its study can help us work out potential directions for global assimilation with newer and often very recent migrations, inward and outward. By this, we mean

DOI: 10.4324/9781003422952-15

that community genealogy projects have relevance for current and future population movements and also social policies.

The significance of the diaspora can vary among different European contexts. Countries such as Ireland, Scotland (Tottola, 2018), Italy, Greece and Poland, with significant historical emigration, strongly promote tourism initiatives targeting their diasporas (Higginbotham, 2012; Li et al., 2019). In Ireland, the Epic Emigration Museum in Dublin City has developed as a national Diaspora centre working as a cultural tourism initiative aiming to add up to 300,000 visitors per annum, as well as aiming to provide a focal point for an otherwise fragmented tourism offering (Miley, 2013). The size of the Irish diaspora is difficult to measure but published figures range from 60 to 80 million (Boyle et al., 2013) with approximately 30 million claiming Irish ancestry in the USA in 2022.[2] Most of those people are more than second-generation Irish, unlike Britain where most people with Irish ancestry are first- or second-generation migrants, with 10% of the current British population having an Irish parent or grandparent. It is believed that up to 14% of Canadian[3] citizens claim full or partial Irish ancestry, while just under 10% of Australians[4] do so also. Such a large global diaspora, dating back 300 years, has a significant impact on tourism figures with approximately 30% of overseas visitors to Ireland claiming ancestral ties. The number of overseas tourists visiting Ireland has increased from approximately 8 million in 2007 to just under 11 million in 2019, i.e. pre-Covid, while numbers had recovered to 7 million in 2022[5] following pandemic disruptions. Tourism metrics are tracked by Fáilte Ireland, the National Tourism Development Authority, and currently the most common method for measuring tourism activity at local level is to gather room and bed-space occupancy data from the full range of accommodation providers.

While it is difficult to determine the economic value of cultural tourism in Ireland, a related figure of €1.5 billion per annum has been suggested by the statutory body, the Heritage Council, which commissioned in 2011 a study of the economic value of the Irish historic environment (ECORYS, 2012). Admittedly, the historic environment (built and natural) attracts visitors from both the New World diaspora populations listed above but also from the large contingent of mid-European tourists who visit from Germany, France and Italy in particular, without known ancestral links.

## 15.1   The historic graves project in Ireland

The Historic Graves Project (Toscano, 2019; Toscano & Tierney, 2020), based in Ireland, uses geolocated community genealogy surveys of historic graveyards to generate a grassroots tourism product in which every parish with a graveyard becomes a tourism resource. Starting in 2011, the web platform has published over 900 hyperlocal[6] gravestone surveys containing historic biographical details for over 250,000 individuals across 15 counties. Recognised as a nationally significant genealogical tourism resource since 2015 (Kennedy & Lyes, 2015), the project has seen its main market shift from a local base (Ireland and the UK) to a global audience, based predominantly on the Irish diaspora populations in Australia, Canada, New Zealand, the UK and the USA.

A community-led approach was built into the Historic Graves Project from the very beginning. Designed and built by archaeologists, it was initially intended to be a community archaeology project. Community archaeology is a relatively recent phenomenon in Ireland and our own engagement with it dates to the World Archaeology Congress (WAC6), held in UCD in 2008. At WAC 6[7] several different community archaeology models were evident, and the main contrast was between the English community archaeology approaches (Thomas, 2014) and a series of indigenous community projects shown from North America and Australia, in particular. Neither model fit our Irish experience and WAC 6 stimulated an internal dialogue within the team about defining possible archaeology projects that would have equal positive benefits for communities as well as for archaeologists. Rural Irish communities have a strong community voluntary ethos, particularly in local sporting organisations, and lessons learned in this context greatly informed our approach to designing community archaeology. Community volunteering organisations in Ireland rely heavily on local volunteers (their unremunerated time input essentially having a financial value), have limited resources and have low donor potential (Gallo & Duffy, 2016).

Two key lessons we learned from community volunteerism that benefited the Historic Graves Project were:

1 The problem is often the opportunity.
2 Funding will be found if the community agrees on the problem and the solution.

In Irish heritage historic graveyards, care and maintenance were a serious problem for underfunded statutory authorities and for local communities deeply invested in the burial grounds (most of which are recorded archaeological monuments with associated protections) where their own family members were often still being buried. Our proposed approach was informed by local community needs and focussed on survey, recording and online publication (usually within a few weeks rather than months) as a first step to conservation and maintenance. We also brought an organisational ability to assist informal community groups in making funding applications for surveys and subsequent conservation works[8].

Building trust and maintaining it over the last 13 years has been an important part of the project and a key element in our request that communities share ownership of the heritage surveys under Creative Commons licences rather than regular copyright; another element of innovation developed within the digital-born origins of the project.

Digital cameras had become mainstream in field archaeology by 2008–2010. We were no longer using film cameras by then and, allied with affordable GPS loggers developed for hobbyists, the geolocation of digital photographs was now accessible to the public. While not as accurate as differential GPS, the early adoption of this technology offered a new element of local history and genealogy, i.e. the geolocation of genealogical datasets which was perhaps our main heritage tourism innovation back then. Geolocated genealogical datasets are a tourism resource because they become points of interest in maps. In our case, our genealogical documents were carved headstones inside historic graveyards – they were already points of interest in a historic landscape and their online publication would serve two purposes. Firstly, to accommodate online individual family tree research and secondly, potentially, to bring those seeking such information to that place. The latter point was obvious to us as we worked with local communities. Self-selected volunteers with a deep interest in their family histories recorded their own historic graveyards with us, and soon began to independently visit other graveyards and associated communities as the database grew.

Our system mapped the location of graveyards and then geolocated the mortuary monuments (with attached biographical detail) within those graveyards, immediately making the places themselves more accessible and easier to find. And while Irish people led the way, the involvement of diaspora visitors wasn't far behind. Our original survey design was based on Mytum (2000) and our original purpose was to follow his methodology, aiming for a detailed survey combining archaeological and genealogical data. However, at a very early stage, we decided to focus on the genealogy first, to build up a broad geographical database of mortuary monuments and geolocated photographs with biographical details as the main priority rather than the complete archaeological record. This strategic shift resulted in an interesting counterpoint to the Mytum approach in England resulting, we believe, in the building of a first stage 'shotgun' survey, covering a wide area in low-medium detail, in terms of archaeological data, which can subsequently be used for more detailed archaeological purposes.

A further element in the design of the Historic Graves Project was the role of the EU LEADER[9] funding source. Between 2011 and 2013, LEADER was the main funding source for the community surveys. LEADER funds come with a strong community-led ethos which considerably benefited our surveys. Equal partnerships were established between the survey trainers and the local communities, with the local groups bearing the main responsibility for driving and completing the surveys. The Leader funds were managed by Local Action Groups and county councils. Still, not all such pairings had an equal interest in community heritage projects, so the initial intention of building a complete national resource soon changed to a more pragmatic tactic of 'follow the funding'. As a result, some countries are very well surveyed/published and others less so, in particular those already engaged with their own solutions (e.g. Kerry, Galway and Clare).

## 15.2   Surnames, genetics and migrations in community genealogy

At first glance, Ireland seems especially suited to genealogical tourism. Island geopolitics of the 10th and 11th centuries resulted in the development of hereditary surnames, approximately 100 years earlier than in neighbouring England and 500 years earlier than in Wales (King & Jobling, 2009). Indeed, in the 5th–7th centuries, standing stones inscribed with Ogham contained personal names and possible sept affiliations in Early Primitive Irish and Latin (Stifter, 2022).

Interestingly, with the development of large-scale global migration since the mid-1600s, the geographical clustering evident in genetic kinship must now be changing significantly. One genetic study test of being a local was having four grandparents born in the same county or province; with increased global mobility in the last 20 years, the question arises as to whether this method of assessment will be as easy to fulfil for populations in the next 100 years.

As Irish septs were highly localised, surnames can then have strong geographical concentrations. For example, O'Dohertys dominate in the Inishowen peninsula on the north coast and McCarthy's, while being found all over the southern province of Munster, are concentrated in Co. Cork, where one of the senior lines are associated with the famous Blarney Castle and the Blarney Stone.

It is interesting to compare this 10th-century Irish naming system with a still surviving alternative Irish system that matches the medieval Welsh system, which is the use of a tripartite name structure using forenames only e.g. a famous 20th-century matchmaker in Co. Kerry was known locally as Dan Paddy Andy, indicating three generations of forenames being Dan himself, Paddy his father and Andy his grandfather – this system was designed to generate a unique identifier at a local level that could also involve matrilineal naming patterns if that clarified identification, and is especially useful in areas with a limited range of surnames e.g. O'Keeffe in the Clonmeen/Banteer/Kanturk triangle of Duhallow. This naming system is still used in parts of Ireland (Lele, 2009).

Until 100–120 years ago, much of Ireland was bilingual and this manifested itself in official documents, including on gravestones, in several different ways. Firstly, official documents were mostly in Latin or English. Headstones are predominantly in English all over Ireland, even for parts of the country that were still using the Irish language, probably because most Irish graveyards were administered through the Church of Ireland until the 1860s (McDermott, 1970). While surnames were mostly written in their English forms on most gravestones, in some cases, the English spelling is derived from Irish pronunciation e.g. Daly in Irish is *Ó'Dalaigh* which is often pronounced O'Dawly and that form is encountered in some localities in Munster. Phonetic spellings are often encountered before the establishment of a national school system in the 1830s, with names like Wynn being spelt Wing and similarly, Flynn being found as Fling in parts of Co. Waterford.

All of this becomes particularly relevant for genealogical tourism as the global expansion of the Irish diaspora began, particularly throughout the 19th century. In the early 1800s, economic depressions, and increased land agitation, allied with fever epidemics, pushed Irish families to emigrate to parts of Wales, England and Scotland in large numbers, especially in areas associated with the Industrial Revolution, such as south Wales and the milling areas of Lancashire and Yorkshire. Indeed, the first English civil registration of 1837 hints at early 1800s migration being influenced by geographical proximity – migrants from Leinster and the Irish midlands travelling by sea from Dublin to Liverpool and its broader hinterlands of Lancashire and West Yorkshire, while southern Irish surnames appear to be found more in London than in northern England (Redmonds et al., 2011).

Surnames, combined with DNA analysis, are a powerful tool for genealogical tourism, although not a silver bullet in many ways. The real benefits of DNA to genealogy appear to lie on the personal level, especially where emigration split up families who subsequently lost touch. There are also strong personal benefits for adopted individuals who can learn via genealogical DNA sites their country or region of origin as well as family lines. However, the ability to assess population movements across Europe and the extent to which old families clung on in localities or to which new families moved into regions entail a detailed local focus for genealogical tourism, which will then be mirrored by following where migrant families spread.

Until recently, surnames were a keyway of signalling ethnicity and identity. At a local level, names were used to assess social status. While at the diaspora level, many people use surnames to assess ethnicity, with all the cultural elements that might signify, stereotypical or otherwise. The connection between surnames and social status is interesting, especially when a historical perspective is taken.

However, the study of the relationship between surnames and genetics is still in its infancy. Cases have been published indicating linkages with key historical figures as common ancestors, going back 100s of years with one of the first such analyses involving the Guinness family (Guinness, 2008) and even over 1,000 years (e.g. O'Neill & O'Briens), but a more multidisciplinary approach seems warranted (Swift, 2013). It is possible that Irish surnames were generally given to male members of a sept, regardless of lineage. Local polities plotted by Tudor administrators,[10] for example, could be called *Pobail* (people of) (Heffernan, 2022), which is of interest because it allows men to be attributed a surname with the application of new bureaucratic systems, without a genetic or blood connection i.e. the name 'people of the O'Keeffe's' can be attached to families who were tenants but not offspring of a high-status O'Keeffe male.

Many Irish people can trace their family trees to the early 19th century, but it is often difficult to get back into the 1700s, unless higher-status families are involved and, even then, most genealogies older than the 1700s are considered suspect, with each case having to be treated separately on its own merits (Swift, 2013). As Irish gravestones only developed a broader social use from the early 1800s onwards, and a really broad usage from the early 1900s onwards, historic gravestones lose their general applicability in community genealogy before the 1750s, but they have value in establishing solid genealogical foundations from which to work backwards.

Our Historic Graves research model was designed to tap into the strengths of Irish surnames. Gone was the earlier model of recording the gravestones of significant families or individuals[11] or of recording only those records of one's own family – community genealogy recognises the importance of doing all the records and the broader benefits for home and diaspora demographics. Community genealogy is driven by the intrinsic factor that gravestones represent family relationships within a community, and that we are not simply mapping family memorials but also identifying networks of connections between different family lines, fostering the generation of good quality family trees, which are important tools for community genealogy.

A genealogical family tree becomes a tourism or diaspora engagement resource through the addition of location to people's life biographies. Family trees become itineraries for planning tourism visits or even for developing self-identity based on ethnicity. Depending on the smallest size of church or civil administration units used in a country, quite fine granularity can be achieved, or aspired to, when making genealogy local. Hyperlocal worked for us in two ways – firstly as an artefact of digital marketing and communications (as text became hypertext with HTML) and secondly by our ability to assign genealogical data to very precise locations, i.e. from a 20m wide bubble of space in a graveyard, using low-cost GPS devices a decade ago, to less than 5m in 2023, using smartphone GPS technology. This means that digital heritage practices can bring genealogical tourists not just to the smallest administrative land unit, e.g. a village or a townland (average 70 acres in size) but often to precise places within those administrative units, meaning genealogical tourism resources are no longer centralised in archives and museum but are found throughout the country. In Ireland, historic land administration units vary through time, but they do share a common history with surname development. Family names are often specifically associated with particular baronies, civil parishes, church parishes and townlands within, with most Irish genetic clusters relating to provincial boundaries (Gilbert et al., 2017). The Irish townland is a subdivision of the larger parish, which when combined with a range of historic (OS maps) and genealogical records (census, valuations, burial records for example) can make for a very precise tourism offering (see challenges of the Peter Robinson case study below).

In Ireland, some surnames are so frequently encountered that one can meet individuals who claim to share a surname for all four grandparents without consanguine relationships, e.g. the Ryan surname is frequently encountered in Co. Tipperary and there are 600 different identified nicknames applied to differentiate one branch from another. With such geographic concentrations of surnames and with the challenges of increased migration into European countries, it is no surprise that surnames and claimed ethnicity have become politically charged. In recent years, anti-immigrant campaigns often use the anti-immigrant phraseology of 'Ireland is Full' and 'Ireland for the Irish' and surnames seem to be one method for defining Irishness.

If one agrees that citizens of a democratic country have equal status and that countries have moral and legal responsibility for fair and decent treatment of refugees and migrants, as well as the potential to benefit from immigrant families, then the negative use of ethnicity must be considered in community genealogy projects. Are we claiming that we are identifying the core local families who define a place over many generations or are we rather identifying a dynamism of occupation involving many families, many of whom do not persist for long at all? Taking a longer time perspective, we think this question answers itself quite quickly. Both archaeological[12] and sociological studies (Roth & Ivemark, 2018) view ethnicity as having strong self-selection elements.

As for North America, the development of the Irish diaspora is dominated by two separate but long-running phases. Firstly, the Protestant/Presbyterian Scots-Irish who had migrated to Northern Ireland following the Ulster Plantations (1606 onwards), subsequently migrated to North America in large numbers throughout the 18th century (Horning, 2002), becoming a significant proportion (around 30%) of the population of the Appalachia region in particular. Secondly, the number of Irish migrants to North America increased considerably throughout the 19th century (Ó Gráda, 1977), partly explained by the massive growth of the Irish population from a total of 5 million in 1800 to over 8 million in the 1840s (Ó Gráda, 2019), with particular increases at the time of the Great Irish Famine (1845–1852) (Crowley et al., 2012).

It is estimated that over 250,000 Scots-Irish left Ireland for America in the 1700s, followed by between 2 and 4 million Irish emigrants throughout the 1800s. These broad movements of people within the then British Empire were largely anonymous until recent times and the increased ability to give names to the emigrants has its roots in the digital genealogical revolution mentioned earlier, providing the potential for good quality historical investigations.

## 15.3    Assisted emigrations

In contrast to the general waves of emigration experienced by the Irish, several specific assisted emigration schemes are very time and place-specific. The politics and practicalities of assisted emigrations were much discussed throughout the 19th century within the British Empire, and it appears local resistance prevailed except in times of crisis. Two of the better-known Great Irish Famine-assisted emigrations were the Landsdowne Estate Assisted Emigration scheme, which sent over 234 South Kerry people to Manhattan in New York, and the Grey's Orphan Scheme, which sent 4114 orphan girls from Poor Relief Workhouses in all 32 counties to Australia in 1849 and 1850 (Ó Gráda, 2019). Another well-known assisted emigration was organised by the Quaker James Hack Tuke (Moran, 2022) and was focused on a localised famine in Connemara, in the late 1870s and early 1880s. Tuke raised substantial charitable contributions in England, which were then used to pay for the assisted emigration of 1300 people from the Clifden area of Connemara to the USA in 1882.

### 15.3.1   Peter Robinson–assisted emigration: case study

The Peter Robinson–Assisted Emigration[13] (PRAE) (Slattery et al., 2016; Donnelly 2009) was very specific in time and people. Two ships carrying almost 600 migrants (or economic refugees) left Cobh/Queenstown in 1823 and a further nine ships in 1825 bringing almost 2000 named people. It was organised by a former soldier, fur traders and colonial administrator Peter Robinson, a British Loyalist who left the US during the Revolutionary War, settling in the Kawartha Lakes area north of Toronto, Canada. For genealogical tourism purposes, it is interesting to note that approximately 307 families are represented, being selected by three key landlords and another 37 smaller landowners. The 1823 group consisted of 182 families, and they principally occupied lands in the Ottawa Valley. The second larger group then arrived in 1825 and were allowed 70-acre farm lots on the Upper Canada frontier lands, in the Kawartha Lakes area where lands had only recently been 'acquired' in an 1818 treaty signed with Kawartha Mississauga First Nations (Pammett, 1936), which was going to pay the equivalent of 10 dollars per annum to the members of the local tribes but only for the lifetime of the living First Nations people, for an area amounting to approximately 2 million acres. Having the biographical details of about 2,500 people who moved from the Blackwater Valley in Ireland to the Ottawa Valley and Kawartha Lakes in Canada has diaspora significance for both sides of the migration.

Our involvement in the project (a Leader project led by Ballyhoura Development CLG) has primarily focussed on the survey of as many historic graveyards in the region from which the families emigrated, in order to build up a database which could potentially bring related diaspora families to their 'home' places in Ireland. Allied with over 100 graveyard surveys published on the Historic Graves platform (Figure 15.1), a systematic attempt was made by medieval historian and genealogist Dr Paul MacCotter (MacCotter, 2013) to trace the point of origin for the families listed on the passenger lists for the PRAE, using the conventional genealogical information sources associated with birth, deaths and marriages. On the Irish side of the equation, Dr MacCotter summarised that approximately 10% of all families could be traced to the townland of origin, with 25% of families being traced in the admittedly scant church records of 1800–1830 (Slattery et al., 2016). Combined with the surname patterns discussed earlier, approximately 40% of the Peter Robinson settlers are considered traceable to relatively precise locations.

*Figure 15.1* Local community and Eacthra team surveying the Old Graveyard in Kilbehenny, as part of the Ballyhoura PRAE Project. Photograph by Hannah McMahon.

On the Canadian and United States side, some patterns have emerged. Firstly, some families have remained on or very near the farms they were granted. Secondly, intermarriage between the families commenced on the three-month sea journey over and continued for generations afterwards. Third, many of the families generated lineages who moved on from the core Canadian territories settling throughout Canada and the USA, and that, because of the detailed biographical data available for family groups, have become highly traceable elements in the otherwise overwhelming Irish migrations to the continent.

An interesting element of the PRAE is the time depth revealed in the colonial process. A core area of the Peter Robinson emigrants was Doneraile in north Cork. This was also the core of the 1580s Tudor plantation of Munster, which resulted in the devastation of local populations and significant changes in the local elites, with the old Gaelic and Old English lordships being largely replaced by new English families. Although temporarily displaced by the later Nine Years War (1594–1603), they were part of a 100-year-long process which led to the establishment of the Protestant Ascendancy by the 1680s, and the various inequalities based therein. The economic migrants of the 1820s PRAE were in many cases the descendants of the survivors of the colonial depredations of 1580–1680, and with their transplantation to Ontario, they were settled into lands farmed and hunted (owned) by First Nations polities.

A community genealogy approach to this type of migration episode can consist of the following exercises. Firstly, we propose historic graveyard surveys as a gateway heritage project, as such surveys are very accessible and a good introduction to other heritage projects, using principles of field recording, transcription, and research. These are then to be followed by geolocated photographic surveys of identifiable farms and homesteads, in both the place of origin and arrival.

Graveyard surveys are easier to do than settlement/farmstead surveys, as burial places are usually more confined and static than settlements and homesteads. Settlements occasionally persist under a single family but more often involve a sequence of structures and associated family groups depending on marriage, intermarriage, and onward migration of the occupying families. Limitations to this approach to community genealogy include copyright, privacy and heritage legislation as well as the availability of resources and sources. For example, legislation in several countries prohibits the publication of private data on gravestones and restricts the publication of sources.[14] Also, the further back we go in time the more the need for professional input into the projects increases. Most community groups can access sources such as the 1901 and 1911 population censuses in Ireland, but localised estate and village records from 1750 to 1850 are often only accessible, decipherable, and meaningful to trained historians.

A significant challenge remains in how to trace the subsequent generations who settled elsewhere in Canada and the USA. As this process continues by individual family tree researchers, the potential for locally significant tourism projects emerges, but good quality family trees are required for this. With the Peter Robinson settlers in mind, we have surveyed 120 graveyards in the core areas of North Cork and South Limerick, creating the background dataset within which the Peter Robinson settlers' surname patterns are found. The impetus of the 200th anniversary of the 1823 and 1825 migrations is leading to a variety of projects and events being coordinated by groups in Canada and Ireland. When the anniversary has passed, we believe the real benefit will continue as the citizen-science/community genealogy works continue to accumulate good quality genealogical datasets, of benefit not just to direct descendants but to other members of related communities also. Shared effort and wide benefits create positive engagement, and this is a key element of this case study.

## 15.4   Communities measuring tourism

The final element we propose for a community genealogy project is to build a broader tourism product, based on the communities' researches but designed to develop communications with the genealogical tourists at several levels. As part of the INCULTUM project, we tested two connected tourism products – the first is an online destinations module, based on a genealogical approach and built into the Historic Graves website. The Destinations module[15] contains a geographically defined region with associated genealogical

*Figure 15.2* A pocket-sized trail for eight POIs in Ardmore roundtower graveyard.
*Source:* John Tierney.

stories, for example the Ballyhoura region, and links to the early 18th century Palatine refugees, who travelled to Ireland from the Rhineland[16]; many stayed in Ireland, but others migrated from the 18th century onwards to the USA, including at least two families who participated in the Peter Robinson emigration (Youngs and Teskey families).

The second genealogical tourism product is an A7-size brochure (Figure 15.2), produced in printed and digital format and branded The Past in Your Pocket. This product is based on a community survey and research into local stories, and is designed to develop communications and gather feedback from heritage tourists. The zig-zag accordion file includes a cover, a map, acknowledgements, background information, a QR code with a tourism questionnaire, and eight points of interest, each one with a photograph and a QR code linking to a web page or digital asset; designed to be used with the A7 printed brochure in one hand and triggering interactions using one's smartphone in the other hand. The digital version of the brochure is a simple PDF file with hyperlinks and can be accessed via a download link in a local tourist office or from a dynamic QR code affixed to a sign on the graveyard gate. A trial of this system was used in one of our community group graveyards (Ardmore, Co. Waterford[17]) and a surprising amount of tourism information was garnered.

Firstly, the tourism questionnaire was triggered on a sign fixed to the gate. While maintaining user anonymity, we were able to gather information on the time and date of the visit as well as the town/region/country of origin of the visitors. Using the QR codes for each of the eight POIs (point of interest) we were able to measure the number of clicks and assess the level of interest from visitors for each of the POIs and also consider site characteristics that might favour clicking one POI and not another; for example, sheltered indoor POIs did garner more QR scans than outdoors (see Table 15.1 for QR scan metrics associated with an Early Medieval ogham stone situated inside the ruined cathedral building in Ardmore graveyard). An attractive element of this product is the minimal investment required in onsite signage, as well as following good practice in heritage signage (ICOMOS, 2008) resulting in low cost, clear and non-intrusive interpretation. The A7 POI pages can also be printed as cards, laminated, and affixed adjacent to the POI. As part of the INCULTUM experiment, to assess differences between usage of the printed and digital products, we measured the use of both the brochure and site-specific cards. Forty printed A7 brochures were issued to attendees of a site tour at the end of July 2023, while the POI cards were affixed from early July onwards.

*Table 15.1* Example of number and percentage of scans/clicks of the QR codes, according to the provenance, for Point of Interest 02 (an Early Medieval Ogham stone) in Ardmore graveyard, Co. Waterford

| Rank | Country | Scans/Clicks | % of Scans/Clicks |
|---|---|---|---|
| 1 | Ireland | 245 | 52.46 |
| 2 | United Kingdom | 60 | 12.85 |
| 3 | United States | 42 | 8.99 |
| 4 | Germany | 26 | 5.57 |
| 5 | France | 21 | 4.5 |
| 6 | Spain | 14 | 3 |
| 7 | Netherlands | 13 | 2.78 |
| 8 | Italy | 12 | 2.57 |
| 9 | Poland | 6 | 1.28 |
| 10 | Australia | 4 | 0.86 |
| 11 | Austria | 4 | 0.86 |
| 12 | Switzerland | 4 | 0.86 |
| 13 | Belgium | 3 | 0.64 |
| 14 | Romania | 2 | 0.43 |
| 15 | United Arab Emirates | 1 | 0.21 |
| 16 | Czech Republic | 1 | 0.21 |
| 17 | Colombia | 1 | 0.21 |
| 18 | Hungary | 1 | 0.21 |
| 19 | Iceland | 1 | 0.21 |
| 20 | Portugal | 1 | 0.21 |
| 21 | Malta | 1 | 0.21 |
| 22 | Denmark | 1 | 0.21 |
| 23 | Canada | 1 | 0.21 |
| 24 | Luxembourg | 1 | 0.21 |
| 25 | Sweden | 1 | 0.21 |

## 15.5   Conclusion

Despite a decade-long history, the INCULTUM project represented a unique opportunity for the Historic Graves project to further refine and extend its methodology of community genealogy projects, including innovative tourism-oriented products to complement the previously used set of elements, composed by geolocated photographic graveyards surveys and field surveys, surnames studies and family trees.

As seen, recent innovations in the community genealogy approach on the Historic Graves web platform (historicgraves.com) include a tourism module that identifies local communities as Destinations and creates stories, itineraries and trails as appropriate, including a small format printed brochure, which uses dynamic QR technology to extend the geolocated heritage database into a means of interacting with and capturing tourism related intelligence such as peak visits, and identification of origin/tourism markets, potentially augmenting the conventional tourism data being gathered already.

Our original intention was to communicate with tourists via the onsite questionnaire only, however, as only a small percentage of site visitors participate in such questionnaires, we believe there is merit in augmenting the data using the dynamic QR codes on the POI cards, as these capture a complimentary dataset, i.e. users choose to interact with a QR code to trigger information to be delivered to them. This approach can also measure activity levels on the site and will allow comparisons to be made between sites. The first experiment focused on a site-specific graveyard trail, which was selected with the intent to test a less niche trail, offering information on the major heritage/tourist points of interest in the broader parish. The model is adaptable and scalable to the nature of the local heritage resources and the preferences of the local community researchers.

The Past in Your Pocket brochure provides heritage information to visitors, but it also garners local tourism intelligence otherwise not collected, without requiring additional efforts from the tourist. Voluntary community surveys generate hyperlocal genealogical datasets that help increase a sense of place, attract tourists and now have the potential to generate hyperlocal tourism intelligence, augmenting the more established Fáilte Ireland methods of measuring bed-night metrics.[18] This then places local communities in the role of collecting tourism data that can be of use for local, regional and national strategies, particularly in planning for the management of tourist flow to less visited areas.

The lessons learned from this Irish case study are relevant to other European countries interested in developing their genealogical tourism resources. Genealogical tourism organised on a community-led basis is a pragmatic model for crowd-sourced digital humanities projects. Even if local legislation does not suit the publication of grave memorials online, other genealogical resources can be substituted for gravestones. Genealogical tourism is not strictly seasonal in nature and therefore has the potential to extend the length of the tourism season, particularly for rural areas.

## Notes

1 https://www.irelandxo.com.
2 https://data.census.gov/table?q=B04006:+PEOPLE+REPORTING+ANCESTRY.
3 Statistics Canada, 2016 Census of Population, Statistics Canada Catalogue no. 98-400-X2016187, www12.statcan.gc.ca. Retrieved 18 February 2024.
4 "Census of Population and Housing: Cultural diversity data summary, 2021". Abs.gov.au. Retrieved 18 February 2024.
5 https://www.itic.ie/RECOVERY/competitiveness-2023/; Covid related travel restrictions were mostly lifted in Ireland in January 2022.
6 https://en.wikipedia.org/wiki/Hyperlocal.
7 https://www.ucd.ie/wac-6/.
8 For example, Templebreedy SOS group in Co. Cork received grant funding three years in succession from 2020 to 2022 for survey and conservation works.
9 https://ec.europa.eu/enrd/leader-clld_en.html.
10 Heffernan and Gleeson at https://youtu.be/zEkDOWKSvJ8?si=y5BjE6PtOD6_i5un&t=2273.
11 Like in the Journal of the Association for the Preservation of the Memorials of the Dead in Ireland the https://www.irishfamilyhistorycentre.com/pdf/?product_id=6262.
12 Carlin in https://youtu.be/i4I52UNkPpE?si=nMduV2GR296f-GBD&t=3634.
13 https://historicgraves.com/project/graveyards-ballyhoura-peter-robinson-assisted-emigration-project.
14 See https://www.ireland.anglican.org/parish-resources/504/guidelines-concerning-parish-records-memorials for some information on the Irish situation.
15 https://historicgraves.com/destination/ballyhoura.
16 See https://historicgraves.com/story/ballyhoura-palatines-german-colony-south-limerick and https://historicgraves.com/destination/ballyhoura/palatine-trail.
17 https://historicgraves.com/blog/places/ardmore-graveyard-trail.
18 https://www.failteireland.ie/FailteIreland/media/WebsiteStructure/Documents/Publications/failte-ireland-hotel-survey-research-oct-2022.pdf.

## Bibliography

Boyle, M., Kitchin, R., & Ancien, D. (2013). Ireland's diaspora strategy: Diaspora for development? In M. Gilmartin & A. White (Eds.), *Migrations: Ireland in a Global World* (pp. 80–97). Manchester: Manchester University Press. https://www.jstor.org/stable/j.ctvnb7qvz.11.

Crowley, J., Smyth, W. J., Murphy, M., & Roche, C. (2012). *Atlas of the Great Irish Famine*. Washington Square, NY: New York University Press.

Donnelly Jr, J. S. (2009). *Captain Rock: The Irish Agrarian Rebellion of 1821–1824*. Madison: University of Wisconsin Press.

ECORYS. (2012). *Economic Value of Ireland's Historic Environment*. Final report to the Heritage Council. Dublin: Heritage Council.

Fitzgerald, P. (1999). The Scotch-Irish & the eighteenth-century Irish diaspora. *History Ireland*, 7(3), 37–41.

Gallo, M., & Duffy, L. (2016). The rural giving difference? Volunteering as philanthropy in an Irish community. *Journal of Rural and Community Development*, 11(1), 1–15.

Gilbert, E., O'Reilly, S., Merrigan, M., McGettigan, D., Molloy, A. M., Brody, L. C., Bodmer, W. F., Hutnik, K., Ennis, S., Lawson, D. J., Wilson, J. F., & Cavalleri, G. L. (2017). The Irish DNA Atlas: Revealing fine-scale population structure and history within Ireland. *Scientific Reports*, 7(1), 1–11. https://doi.org/10.1038/s41598-017-17124-4

Guinness, P. (2008). *Arthur's Round: The Life and Times of Brewing Legend Arthur Guinness*. London: Peter Owen Publishers.

Heffernan, D. (2022). *Early Modern Duhallow, c.1534–1641*. Dublin, Ireland: Four Courts Press.

Higginbotham, G. (2012). Seeking roots and tracing lineages: constructing a framework of reference for roots and genealogical tourism. *Journal of Heritage Tourism*, 7(3), 189–203. https://doi.org/10.1080/1743873x.2012.669765

Horning, A. J. (2002). Myth, migration, and material culture: Archaeology and the ulster influence on appalachia. *Historical Archaeology*, 36(4), 129–149. https://doi.org/10.1007/bf03374373

ICOMOS. (2008). *The ICOMOS Charter for the Interpretation and Presentation of Cultural Heritage.* https://www.icomos.org/images/DOCUMENTS/Charters/interpretation_e.pdf

Kennedy, L., & Lyes, M. (2015). *Local Diaspora Toolkit*. Dublin: Office of the Minister for Diaspora Affairs.

King, T., & Jobling, M. A. (2009). What's in a name? Y chromosomes, surnames and the genetic genealogy revolution. *Trends in Genetics*, 25(8), 351–360. https://doi.org/10.1016/j.tig.2009.06.003

Lele, V. (2009). "It's not really a nickname, it's a method": Local names, state intimates, and kinship register in the Irish *Gaeltacht*. *Journal of Linguistic Anthropology*, 19(1), 101–116. https://doi:10.1111/j.1548-1395.2009.01021.x

Li, T. E., McKercher, B., & Chan, E. T. H. (2019). Towards a conceptual framework for diaspora tourism. *Current Issues in Tourism*, 23(17), 2109–2126. https://doi.org/10.1080/13683500.2019.1634013

MacCotter, P. (2013). The M7 motorway historical landscape: Studies in the history of Ikerrin and Elyocarroll. *Tipperary Historical Journal*, 2013, 25–57.

McDermott, R. P. (1970). The church of Ireland since disestablishment. *Theology*, 73(599), 208–214.

Miley, J. (2013). *The Gathering Final Report*. https://www.failteireland.ie/FailteIreland/media/WebsiteStructure/Documents/eZine/TheGathering_FinalReport_JimMiley_December2013.pdf

Moran, G. (2022). From poverty to posterity: The Tuke Committee– Assisted emigrants in Minnesota, 1882–1884. *New Hibernia Review*, 26(3), 34–52. https://doi.org/10.1353/nhr.2022.0031

Mytum, H. (2000). *Recording and Analysing Graveyards, 15*. York: Council for British Archaeology.

Ó Gráda, C. (1977). Some aspects of nineteenth-century Irish emigration. In L. M. Cullen & T. C. Smout (Eds.), *Comparative Aspects of Scottish and Irish Economic and Social History, 1600–1900* (pp. 65–73). John Donald, Edinburgh: Birlinn Ltd. https://hdl.handle.net/10197/388

Ó Gráda, C. (2019). The next world and the new world: Relief, migration, and the Great Irish famine. *The Journal of Economic History*, 79(2), 319–355. https://doi.org/10.1017/s002205071900010x

Pammett, H. T. (1936). Assisted emigration from Ireland to Upper Canada under Peter Robinson. *Papers and Records of the Ontario Historical Society*, 31, 187–188.

Redmonds, G., King, T., & Hey, D. (2011). *Surnames, DNA, and Family History*. New York: Oxford University Press.

Roth, W. D., & Ivemark, B. (2018). Genetic options: The impact of genetic ancestry testing on consumers' racial and ethnic identities. *American Journal of Sociology*, *124*(1), 150–184. https://doi.org/10.1086/697487

Slattery, A., Tierney, J., & MacCotter, P. (2016). The Ballyhoura Peter Robinson assisted emigration project. *Zenodo*. https://doi.org/10.5281/zenodo.10514996

Stifter, D. (2022). *Ogam. Language|Writing|Epigraphy*. AELAW 10. Zaragoza: Prensas de la Universidad de Zaragoza.

Swift, C. (2013). Interlaced scholarship: Genealogies and genetics in twenty-first-century Ireland. In Seán Duffy (Ed.), *Princes, Prelates and Poets in Medieval Ireland* (pp. 19–31). Dublin: Four Courts Press.

Thomas, S. (2014). Making archaeological heritage accessible in Great Britain: Enter community archaeology. In S. Thomas & J. Lea (Eds.), *Public Participation in Archaeology* (NED-New edition, vol. 15, pp. 23–34). Woodbridge: Boydell & Brewer. https://www.jstor.org/stable/10.7722/j.ctt5vj829.9

Toscano, M. (2019). Crowdsourcing Irish history. *3rd National Conference of the Italian Association of Public History – AIPH*, University of Campania Vanvitelli (Santa Maria Capua Vetere, Caserta) 24th–28th of June 2019. https://hdl.handle.net/10481/56247

Toscano, M., & Tierney, J. (2020). Historic graves. *Visualizing Objects, Places, and Spaces: A Digital Project Handbook*. https://doi.org/10.21428/51bee781.4c0eec46

Tottola, M. (2018). *Nuovi scenari di turismo culturale: il turismo genealogico* (thesis).

# Index

Note: **Bold** page numbers refer to tables; *italic* page numbers refer to figures and page numbers followed by "n" denote endnotes.